The Significance of Trees

An archaeological perspective

Peter Skoglund

BAR International Series 2324
2012

Published in 2016 by
BAR Publishing, Oxford

BAR International Series 2324

The Significance of Trees

ISBN 978 1 4073 0907 1

© P Skoglund and the Publisher 2012

The author's moral rights under the 1988 UK Copyright,
Designs and Patents Act are hereby expressly asserted.

All rights reserved. No part of this work may be copied, reproduced, stored,
sold, distributed, scanned, saved in any form of digital format or transmitted
in any form digitally, without the written permission of the Publisher.

BAR Publishing is the trading name of British Archaeological Reports (Oxford) Ltd.
British Archaeological Reports was first incorporated in 1974 to publish the BAR
Series, International and British. In 1992 Hadrian Books Ltd became part of the BAR
group. This volume was originally published by Archaeopress in conjunction with
British Archaeological Reports (Oxford) Ltd / Hadrian Books Ltd, the Series principal
publisher, in 2012. This present volume is published by BAR Publishing, 2016.

Printed in England

PUBLISHING

BAR titles are available from:

         BAR Publishing
         122 Banbury Rd, Oxford, OX2 7BP, UK
EMAIL   info@barpublishing.com
PHONE +44 (0)1865 310431
FAX   +44 (0)1865 316916
         www.barpublishing.com

Contents

List of Figures iii

List of Tables v

Foreword vi

Chapter 1.
INTRODUCTION 1

The archaeological contribution 1

Forms of wood preservation and representation 4

From technical solutions to social biographies 6

Chapter 2.
THE SIGNIFICANCE OF INDIVIDUAL TREES 8

The ritual significance of trees 8

The pictorial evidence – promises and obstacles 10

Spruce or deciduous trees? 11

The pruning of trees 12

Rituals related to trees in the landscape 14

The conductor of tree rituals 18

Trees as mediators between earth and heaven 20

Trees and life-cycle rituals 21

Trees and creation myths 22

A long-lasting practice 25

Chapter 3.
THE ORDERING OF TREES 26

Charcoal – records from bogs 27

Charcoal – records from archaeological contexts 27

The use of trees in the Central Småland Uplands 28

The use of trees in the Western Småland Uplands 32

The structured use of trees 33

The life cycle of trees and cereals 34

Landscape history and the management of trees 34

Trees and rituals 36

Chapter 4.
THE USE AND REUSE OF TIMBER 38

The life cycle of wood and timber 40

Reading wood into house plans 43

The reuse of timber as indicated by house plans 48

A wider perspective 52

Chapter 5.
LINNAEUS' LANDSCAPE 54

The logistics of landscapes 54

Never-ending landscape change 55

The social significance of trees 56

References 58

List of Figures

Figure 1. A row of willows on the border between a field and a road outside Malmö, southern Sweden 3

Figure 2. Pollarded oak (Quercus petrea), district of Riano, Spain 3

Figure 3. Pollarded trees, district of Botiza, Rumania 3

Figure 4. a) A rock-art tree from Kyrkorud, Tanum, b) a rock-art tree from Ör, Småland 12

Figure 5. Human at the top of a tree. Tanum; Bohuslän, RAÄ 66 12

Figure 6. Sketch explaining different ways of pruning trees 12

Figure 7. Map showing the location of northern Bohuslän in Sweden 13

Figure 8. Runohällen with measured terrain curves indicating when the cliff rose out of the sea 15

Figure 9. Detail of Runohällen 16

Figure 10. Leaf-stack on a photo from 1931 from Slätthög Parish, Småland Uplands 16

Figure 11. Part of the panel at Aspeberget 17

Figure 12. The outcrop of 1000 cup-marks 18

Figure 13. The cyclical journey of the sun inferred from motifs on a Danish metal razors from the Late Bronze Age 19

Figure 14. Comparisons between mushroom-shaped symbols on a Danish metal razor, a rock carving from Bohuslän and a cult axe 20

Figure 15. Four men bearing cult-axes at a rock-art site in Simris, Scania 21

Figure 16. Razor from south Jutland with an elaborate tree in a central position aboard the engraved ship 21

Figure 17. Mushroom-shaped symbols on the back of a spectacle fibula from Repplinge, Öland 22

Figure 18. Tree on the back of a spectacle fibula from Repplinge 22

Figure 19. Figure on the back of a spectacle fibula from Kareby, Bohuslän, ending in three branches 22

Figure 20. Razor probably from Lolland, Denmark. At the prow there is a stylized animal's head, probably a horse, close to a mushroom-shaped symbol 23

Figure 21. Hanging bowl from Öland, south-east Sweden 24

Figure 22. Bronze sheet from northern Germany 25

Figure 23. Map showing the location of the Western and Central Småland Uplands in Sweden 29

Figure 24. Clearance cairn, Tjureda Parish, Central Småland Uplands 30

Figure 25. A monumental burial cairn, Tjureda Parish, Central Småland Uplands 35

Figure 26. Plan and section of a reconstructed Iron Age long-house at Lejre, Denmark 39

Figure 27. Diagram showing possible timber use from a tree. A=Lower trunk; B=Upper trunk; HR=High beam; GB=Gable head; VR=Wall head; TS=Roof-bearing posts; TB=Crossbeam; DS=Door posts; VS=Wall posts; BB=Connecting beam; HJR=corner head; DO=Door head; DT=Doorstep 42

Figure 28. Wood from the Priorsløkke site at Horsens, Denmark. To the left is the upper part of a roof-bearing post; in the middle is the bottom of a roof-bearing post; and to the right is a plank, presumably a wall plank 42

Figure 29. Map showing the location of the areas discussed in the text 44

Figure 30. House III at Lockarp with surrounding fences and an outbuilding 45

Figure 31. The inner division of the houses at Lockarp: A=Entrance room; B= Kitchen; C= Living room/Storeroom; D=Barn; E=Threshing barn; F=Stable; G=Kitchen 45

Figure 32. Plan showing the distribution of oak of different age, House III Lockarp. Room division according to Eliasson and Kishonti 2003 and above is the author's interpretation based on the charcoal evidence 47

Figure 33. Plan of House I, House II and House VIII at Lilla Köpinge, Sweden (Trench B14), reproduced with an identical roof-bearing construction. The outer walls are reconstructed from the outer wall posts and the roof-bearing construction is inferred from the roof-bearing posts 51

List of Tables

Table 1. Species-identified charcoal from clearance cairns from Rottnevägen. One sample with a mixture of aspen, birch and oak is not listed in the table. Data from *Smålands museum rapport 2003*: 60 28

Table 2. Species-identified charcoal from the Bronze Age (3430–2340 BP), distributed by context 30

Table 3. Species-identified charcoal from the Early Iron Age (2340–1765 BP) distributed by context 31

Table 4. The dating of clearance cairns with ^{14}C-dated charcoal identified as alder from the Iron Age 31

Table 5. Species-identified charcoal from the Late Iron Age (1760–1005 BP) distributed by context 31

Table 6. Species-identified charcoal from the Middle and Late Bronze Age and the Early Iron Age (3140–1925 BP) distributed by context 32

Table 7. Species-identified charcoal from the Roman Iron Age (1920–1505 BP) distributed by context 32

Table 8. The numbers of ^{14}C-dated pieces of charcoal of hazel wood and charred hazel nuts distributed by context 33

Table 9. Species-identified charcoal from the Late Iron Age (1505–1255 BP) distributed by context 33

Table 10. Number of clearance cairns in different periods and the average number of clearance cairns per 100 years. Based on tables 6, 7 and 9 35

Table 11. Number of clearance cairns in different periods and the average number of clearance cairns per 100 years. Based on tables 2, 3 and 5 35

Table 12. Schematic overview of the social and ritual contexts of various trees 37

Table 13. The room division at House III, Lockarp, Sweden as suggested by the archaeological record, Macrofossil analysis, phosphate analysis and MS analysis. Information from Eliasson and Kishonti 2003: 36–43, 56–61; Eliasson and Kishonti 2007: 208–212 45

Table 14. The number of charcoal pieces from various species at House III, Lockarp, Sweden. Data from Eliasson and Kishonti 2007: 172–173 46

Table 15. The distribution of oak with various ages at House III at Lockarp, Sweden. Data from Eliasson and Kishonti 2007: 172–173 47

Table 16. The distribution of charcoal of oak with various ages from House III, Lockarp Sweden. Data from Eliasson and Kishonti 2007: 172–173 48

Table 17. ^{14}C dates from Houses I, II and VIII in trench B14 at Lilla Köpinge. Data from Tesch 1993: 89 51

Foreword

The idea of writing something on the relationship between man and trees in prehistory emerged during the spring of 2005. At that time my interest was drawn to images depicting trees in Scandinavian rock art. I felt that previous interpretations of these trees as spruces or conifers had many disadvantages – they should rather be seen in the light of anthropological research stressing how man reshapes trees in various ways in order to obtain valuable products.

Slowly but steadily this insight grew into a wider concern and resulted in a question: How can archaeologists contribute to a general understanding of the relationship between people and trees in prehistory?

Thanks to *Carl Stadlers Arkeologiska forskningsfond* I was able to have two months of research leave in 2006. During this period the major ideas discussed in this book were formulated and many of the monographs and articles referred to were consulted.

However, the next years were occupied with other duties connected to my ordinary work in contract archaeology. But the idea of writing something substantial on this topic kept coming back to me. In 2010 I received funding from *Helge Ax:son Johnsons stiftelse*, which financed two months of research leave. During that period this manuscript was finally written.

The manuscript has been read and commented in whole or part by Nils Björhem PhD, Professor Richard Bradley, Professor Joakim Goldhahn and Associate Professor Per Lagerås. I have not always followed their advice; any errors or misunderstandings remaining in the text are my own responsibility.

For some years I closely followed various research projects led by Professor Marie-José Gaillard-Lemdahl, a palaeoecologist at Linnaeus University. The discussions with Marie and her colleagues. Associate Professor Geoffrey Lemdahl, Annica Greisman PhD and Fredrik Olsson PhD have contributed substantially in shaping my view of trees and woodlands in prehistory.

When the basic ideas were formulated I was strongly supported by Associate Professor Eva Svensson. I am deeply grateful for Eva's support in the initial part of this project.

I wish to thank Professor Urban Emanuelsson for help with figures.

Finally, I wish to thank my former and present managers, Martin Hansson, Per Sarnäs, Olof Hermelin and Nils Johansson, for supporting the project by letting me off work.

The writing of this book was made possible by grants from *Helge Ax:son Johnsons stiftelse*, *Carl Stadlers Arkeologiska forskningsfond* and *the Gyllenstierna Krapperup Foundation*.

The English was revised by Alan Crozier.

Väggarp, 10th of September 2011   Peter Skoglund

Chapter 1.
INTRODUCTION

Trees matter perhaps more today than ever. People in the modern world have very personal and intricate relationships to trees. When individual trees are about to be felled, people often react and try to protect and preserve what they regard as important values in the urban or rural landscape. Trees have become important landmarks which materialize the aesthetic and historical dimensions of life. In contrast to buildings and monuments that also fulfil this purpose, trees are fragile living organism that by their very nature inform us about the passing of time, both through their annual cycle and by their ageing and final decomposition (Garner 2004; Konijnendijk 2008).

The fact that trees are important to people is in itself an invitation to the historical sciences to address the cultural history of trees. Furthermore, it has become increasingly obvious in recent years that the combination of history and ecology is important for understanding today's landscapes (Crumley 1994; Kirkby and Watkins 1998; Foster and Aber 2004).

David Foster and his colleagues have written extensively on the importance of studying history in order to comprehend today's woodland. An important conclusion is that the major factor shaping the nature of woods and wildlife in most of the western world is decisions made a couple of generations back in time. This is illustrated by the research carried out at Harvard Forest in central Massachusetts. By combing a wide range of scientific approaches, including the natural sciences and the humanities, an interesting pattern of never-ending change is revealed (Foster and Aber 2004).

When studying the woods of New England in a thousand-year perspective it becomes obvious that there never was a pristine landscape; at the time of European settlement the native people had already changed the character of the wood by various fire regimes, extensive settlement and a mixture of agricultural techniques.

However, the European settlements would affect the landscape of New England in a more far-reaching way than ever before, and around 1830 the former wooded landscape had been transformed into extensive cultivated fields, with only a small amount of scattered trees surviving in between the agricultural plots.

This situation changed dramatically when the former farmers in the mid 19th century in large numbers abandoned their farms to find employment in the growing cities of eastern North America. As the land was abandoned, the older fields reforested and the forest industry expanded. A consequence of the return of the wood was the return of wildlife. Today beer and moose once again inhabit the woods of New England, which was not the case a hundred years ago (Bernandos *et al.* 2004; Foster *et al.* 2004).

Therefore, if we want to understand the woodland of New England, or the woods in any other region that have gone through a similar cycle of deforestation and regeneration, we must turn to history. The cultural and social dimensions of past woodland management are crucial for understanding today's landscape (Crumley 1994; Kirkby and Watkins 1998; Foster and Aber 2004; Hayashida 2005).

THE ARCHAEOLOGICAL CONTRIBUTION

What is the specific contribution of archaeology to this topic? Since archaeology is often concerned with societies that existed several thousand years back in time there are rarely any direct links between the decisions of the prehistoric people studied and today's landscape. The scope of archaeology is somewhat different from what is studied by ecologists and historians – its main contribution is to enrich our understanding of past woodland management and the significance of trees by bringing humans and the long-term history into the discussion (Skolgund and Svensson 2010).

Archaeology is concerned with people. In contrast to, say, palaeoeceology archaeology studies the actions of people in a social setting. Therefore archaeology can broaden the discussion when it comes to understanding how landscape changes as a result of man's actions. Archaeologists study landscape from a broad perspective involving the construction of monuments, roads and houses and the clearing of land for agriculture and grazing. Thereby man's relation to trees and woodland becomes one among many components making up the totality of past cultural landscapes. The character of the archaeological record also enables archaeologists to study the ritual dimensions of man's relation to trees. Understanding the symbolic aspects of trees is important since these cannot be separated from our understanding of the practical use of trees.

Archaeology is concerned with long-term history. From oral traditions, written sources, photographs and pictures we are informed of a broad range of practical and ritual aspects of man's relation to trees (Hæggström 1996; Sillasoo 2006). From medieval church paintings, for example, we can study how people used pollarded trees to cut branches and feed their animals. This way of using trees is common in many traditional societies, and archaeology can add a time-depth to these practices by identifying similar scenes in, for example, rock art. This in turn has consequences for our understanding of today's pollarded trees that still survive in restricted areas of northern Europe – they are the outcome of a tradition going several thousand years back in time.

Archaeology can write a different cultural history. Historians and human geographers approaching woodland history though the lenses of written sources and maps tend to use recent analogies when trying to understand past woodlands. Archaeologists have other sources of inspiration than historians and geographers, and they often work with anthropological analogies. The ideas and analogies of archaeology are neither better nor worse than the working tools of other disciplines, but they are different. Archaeology can thereby contribute to a more pluralistic understanding of humans' use of trees in a historic perspective.

Culturally modified trees

Since the 1980s there has been a growing interest in culturally modified trees. Important studies have been carried out in the Pacific Northeast, northern Fennoscandia and south-eastern Australia. Most studies have dealt with native people and their traditional use of forest, specially the use of bark for fodder, but also blazes found along historically important trails and the use of trees for carvings, along with many other kinds of practices, have been studied (Andersson 2005: 10–20).

A well-known practice carried out in relation to culturally modified trees is the native people's use of cedar trees in the Pacific Northwest. Studies carried out in British Columbia reveal that the native people used the cedar tree and other kinds of trees for a variety of purposes.

> Northwest coast peoples knew the cedar tree intimately, and all of its parts were highly valued. The wood was used to make canoes, paddles, house planks, house posts, crest and mortuary poles, bentwood boxes, bows, masks, bowls and dishes. The fibrous inner bark was fashioned into clothing, hats, mats, masks, rattles, nets, twine, blankets, diapers, towels, and rope. The coarse outer bark was used for roofing material, canoe bailers and canoe covers. The withes, or flexible branchlets that hang down from the main branches, were valued for making heavy-duty rope, fish traps, and baskets. Even the roots were used, to make baskets and cradles (Stryd and Feddema 2005: 5).

The use of the cedar tree and other kinds of trees left a visible impact on the landscape. Still today there remain substantial numbers of culturally modified trees in the Pacific Northwest. Not only the removal of bark but also the removal of planks from standing trees left markers on living trees that are still visible (Stryd and Feddema 2005).

Another example is the collection of bark from pine, a well-documented tradition in the northern hemisphere – in the northern parts of Canada, Russia and Scandinavia. In these areas the bark was one of the very few vitamin C sources that could prevent scurvy during the winter season. Since the bark was taken from a restricted part of the stem the tree survived, and many of these trees are still visible in today's landscape. Their age is impressive; studies of Scandinavian trees reveal that the oldest traces of bark peeling on standing trees are from the 15th century, but by dating subfossil logs we can trace this tradition 3,000 years back in time. These trees are biological by definition but their character is shaped by man – they are on the border between nature and culture (Zachrisson *et al.* 2000; Östlund *et al.* 2002, 2004).

These examples are illustrative, and on a general level the lessons learned from the study of native people's use of trees are relevant also for our understanding of other kinds of cultural landscapes such as today's treatment of trees in an urban setting.

One example could be taken from the town where I work in southernmost Sweden (Figure 1). Walking to work I go through an avenue composed of willows. Today these willows are pruned for aesthetic reasons in the late winter or early spring. However, this practice has its roots in an older tradition where willows were planted in rows on the borders of fields to prevent erosion. In 19th century southernmost Sweden there was a shortage of wood and the willows were cut back in the summertime to maximize the production of branches that could be used for fodder and fuel and for baskets and rafters. The cutting of branches that once was a necessity is today done for visual reasons. Moreover, the pruned willows have become a symbol of identity for the southernmost province of Sweden, symbolizing the open landscape, characterized by individual trees rather than forest, which distinguishes it from many other parts of Sweden.

The pruned broad-leaved trees of temperate Europe that we today recognize as bearers of important values in the

Figure 1. A row of willows on the border between a field and a road outside Malmö, southern Sweden. Photo: Peter Skoglund

Figure 3. Pollarded trees, district of Botiza, Rumania. Photo: Urban Emanuelsson

landscape are the outcome of older agricultural techniques involving the cutting of branches on a regular bases to improve the production of leaves, branches and twigs (Figures 2, 3). To understand this cultural heritage which is embedded in biology we need a historical perspective. Since trees have always been a part of human culture, our studies should not stop when maps and texts cease to exist, but should include archaeology.

Figure 2. Pollarded oak (Quercus petrea), district of Riano, Spain. Photo: Urban Emanuelsson

The examples given above demonstrate that it is useless to define these trees as something that could be studied through natural conservation policies alone – to understand why a planked cedar, a bark-peeled pine or a pruned willow have become what they are today, a historic perspective is necessary. These trees are ancient remains often extending several hundred years back in time. Moreover – as has been stated earlier – we know from archaeology that these trees sometimes are representatives of a tradition going several thousand years back in time.

Archaeology – a discipline of durable materials

How come archaeologists have rarely discussed trees? Most probably the answer to this question is to be sought in the foundation of modern archaeology in the 19th century.

By using a typological approach, archaeologists like the Swedish professor Oscar Montelius were able to sort the archaeological material into chronological sequences. By defining types and investigating which objects occurred together, he was able to estimate which objects were contemporaneous. Sometimes older objects occurred among a younger set of objects and thereby a relative chronology was established. By using imported objects found in northern Europe and linking them to the production places in central and southern Europe a firmer chronology was worked out. Finally, an absolute chronology was established using data from chronologies based on written sources and numismatics (Gräslund 1987).

To be able to define types accurately and to compare objects over vast distances in Europe there was a need for larger collections of objects. In that way archaeology became synonymous with the study of large collections of objects, and the only objects surviving in great quantity were those made of durable materials: metal, stone or pottery.

Of course, the importance of preserved wooden materials was recognized but the study of wood was never integrated in the core of archaeological discourse. Apart from being few in number in comparison to objects made of metal, stone and pottery there was another obstacle with wooden objects; before the introduction of ^{14}C dates wood was very difficult to date. Typology did not work very well since wood handicraft changes slowly and objects made several thousand years ago often resemble those made a couple of generations ago.

Sometimes wooden objects found in bogs were dated using pollen analysis. This was based on a logical assumption: since the change of the wood composition was reflected in the pollen material in the individual layers, an object found in the bog could be related to a particular event in the history of the forest and thus indirectly dated. However, the basis for assigning an absolute date to the introduction of species such as beech and spruce was not very accurate in the early 20th century; this is why many wooden objects were given confusing dates that have been revised thanks to dating methods such as dendrochronology and ^{14}C dating (cf. Skoglund and Lagerås 2002). These problems in dating caused problems and helped to marginalize the study of wood.

Because of these circumstances, wood and trees were rarely discussed in archaeology. Wooden objects were reckoned as interesting but non-essential finds lacking qualities that could make them contribute to a general understanding of prehistory.

However, this situation is not valid everywhere; in the Alps the recognition of wooden objects has followed a different path since we find here larger quantities of wood preserved in the lake settlements. Therefore sites like *Fiavè* are very important since they open our eyes to the importance of wood in prehistory.

The implications of Fiavè

The bog of Fiavè is situated just outside the Giudicaria Valleys, south-west of Trentino in northern Italy. The site in a wide marshy area is famous because of the prehistoric dwelling site that occupied the basin of a lake of post-glacial origin which disappeared because of the progressive growth of peat moss.

The first archaeological findings in the marsh of Fiavè took place during peat digging in the second half of the 19th century. Systematic excavations were carried out by Renato Perini between the years 1969 and 1979. The good state of preservation of perishable materials such as textiles, wood and basketry artefacts, as well as the stratigraphic series of Bronze Age settlements, make this one of the most important peat-bog sites in Italy (Perini 1987, 1995).

Coming to Fiavè today is a strange experience since the posts that were the foundations for the Bronze Age platforms and houses still are visible, as they were deliberately left in position when the archaeological campaign ended in 1979. But what is even more fascinating is glancing through the publications reporting on the structures and finds from the site.

By doing this we get detailed glimpses of life as lived 3,000 years ago. Perhaps most fascinating are the preserved wooden objects. Here we find agricultural items such as ploughs and yokes, compete sickles with both the handle made of wood and the cutting edge made of flint still in place, axes handles and objects connected with the preparation and eating of food, such as whisks, cups, bowls and ladles of wood (Perini 1987).

The wooden material culture of Fiavè resembles very much the objects displayed at museum exhibitions of 19th century folk culture. Fiavè is thus an important reminder that not only the material culture of historic, but also prehistoric Europe, was primarily based on wood. The broad range of wooden object and the excellent craftsmanship applied in making these objects are impressive.

This wood culture could not have existed without a sense of woodland management and an idea of how to use trees for various purposes. These issues are of course best studied at those rare sites like Fiavè where there is a broad range of wooden objects and structural details from the houses and platforms. However, archaeologists are not dependent upon exceptional preservation conditions to study man's relation to wood and trees. There is quite a broad variety of sources informing us about these issues.

FORMS OF WOOD PRESERVATION AND REPRESENTATION

The structure and characteristics of wood and wooden constructions may survive in three different forms: wood can survive in itself, for example, when water-logged; it can carbonize; and when decomposing it may colour the soil, making it possible to identify wooden structures. Moreover trees and structures made of wood might be represented in other media such as pictures, monuments or objects. In the following these different forms for sources to wood and trees will be briefly presented.

Waterlogged wood

Waterlogged wood consists of objects preserved because they have been stored in wet environments which have prevented their decomposition by oxygen. The study of preserved wooden objects and wooden structures has been revitalized by the use of dendrochronology. In much of Europe there are curves for oak going back to the Neolithic, enabling archaeologists to date wood to an exact year or very close to a specific year (cf. Hillam *et al.* 1990).

A special case is preserved wooden ships. These have two different origins: they might be found in the sea, sometimes reused as underwater barriers, like the Viking ships found at Roskilde Fjord in Denmark, or on rare occasions they may survive beneath burial mounds, like the Oseberg ship in southern Norway (Shetelig *et al.* 1917–2006; Crumlin Pedersen and Bondeson 2002). In England plank-built ships are known from the Early Bronze Age (Wright 1990; Clark 2004), while in Scandinavian the oldest preserved ship dates to the very Early Iron Age (Crumlin Pedersen and Trakadas 2004).

Wooden ships are very informative structures enabling us to study wood craftsmanship in detail and how it changed over centuries and millennia (Larsson 2007). However, these constructions were also special in prehistory, and they were never part of a land-based everyday life, which makes it difficult to draw general conclusions from these kinds of finds.

Another special case is some of the Danish Bronze Age mounds – excavated in the late 19th and early 20th century – which contain iron-pan-encapsulated, wet mound cores, in which organic material has been extraordinarily well preserved. Recent studies and experiments seem to demonstrate that these extraordinary preservation conditions are the outcome of a burial ritual where an original and smaller mound was flooded with water and after some time covered by a larger mound (Breuning-Madsen et al. 2001; Breuning-Madsen et al. 2003). These graves have broadened our view of Scandinavian wooden material culture, since not only personal items were preserved in the oak cists, but also small wooden objects and in a unique case a preserved wooden folded chair (Boye 1896; Randsborg and Christensen 2006).

Another category is structures that were partly built to function in wet environments, such as roads crossing marshes and boggy areas. Because of this, archaeologists know quite a lot of ancient roads constructed of wood, and by analysing the trees used, it has been possible to draw conclusions about the kind of timber used in these structures (Coles and Coles 1986; Hillam et al. 1990).

Apart from special sites like the Alpine lake settlements or the Irish crannogs, wooden objects connected to an ordinary land-based life are primarily found in bogs, rivers or lakes and not at settlements. Because of this, many wooden objects have been found randomly, without archaeologists being at hand, during peat cutting or construction work. The majority of finds came to light during the intensive peat-cutting periods during the first half of the 20th century (Schovsbo 1987: 41–42). Wagons, agricultural implements and objects connected to food preparation are among those objects found in these contexts.

Recently this situation has partly changed; as a consequence of large-scale excavations of settlements wells have become an established category of features at archaeological excavations. Because the risk of walls collapsing, these features are difficult to excavate but since they are wet environments they have the potential to store well-preserved wooden materials. The wood might be part of the well construction itself or it might be secondary refuse or ritual deposits informing us about how wood was used in other contexts at the settlement.

Carbonized wood

Finds of charcoal macro-remains from Swiss lake settlements were studied back in the 1860s, but it is only recently that the sampling of charcoal on archaeological sites has become a standard method. This is due to the breakthrough of radiocarbon dating in the 1960s and the further development of the technique in the preceding decades.

Today AMS (Accelerator Mass Spectrometry) radiocarbon dating is a standard dating method. Samples that are far tinier than that needed for standard radiocarbon dating can thus be used. Standard ^{14}C dates require amounts of between 1 and 10 grams of charcoal; AMS can use as little as 1–2 milligrams.

The study of charcoal remains is a discipline of its own studying topics such as vegetation history and fuel selection. However, the breakthrough of ^{14}C dating has made the sampling and interpretation of charcoal and ^{14}C dating into an integral part of archaeological practice.

To be able to pick out the finest charcoal pieces for dating, the tree species of charcoal from archaeological sites is regularly identified and often the age is estimated too (Bartholin and Berglund 1992). By choosing wood with a low age one will reduce the way the specific age of the wood affects the radiocarbon result.

Because charcoal routinely undergoes species identification and age estimation, there is today a huge amount of carbonized ^{14}C-dated wood from archaeological contexts. This material was not primarily sampled to answer questions related to trees and woodland management, but it is in fact excellent for these purposes since every piece of charcoal is dated by ^{14}C. A difficulty is that a single excavation often produces a limited amount of charcoal identified to species. However, in larger projects this kind of information has been successfully used in interpretations regarding the handling of trees in prehistory (Lagerås 2000; Lagerås and Bartholin 2003).

Wood as colourings

When wood ends up in the ground, decaying colourings may occur in the soil that can be interpreted in meaningful ways. In quantitative terms, house remains dominate the picture. By interpreting the colourings left by the roof-bearing posts it is possible to reconstruct the basic layout of a house.

In principle all kinds of wooden structures put into the soil can leave identifiable traces when rotted away. For example, the Danish Iron Age grave-field of *Slusegård*, on the Island of Bornholm in the Baltic, has contributed in a substantial way to our understanding of prehistoric boats, even though not a single piece of wood from the boats was preserved. These boats were used as containers for dead bodies, but their main value is that they give detail information about the broad boat types used by ordinary people in the south Baltic about 2,000 years ago (Andersen et al. 1991).

Trees and wooden structures on pictures

Trees and also wooden structures occur in prehistoric pictures. Possibly the oldest recognizable trees in northern Europe are to be found on Scandinavian rock carvings dating to the Bronze Age, 1700–500 BC. In Scandinavian rock art, as in the rock art produced in the southern Alps, we also found structures made of wood. In both regions we find wagons and ploughs, while ships and houses are restricted to Scandinavia and the southern Alps respectively (Malmer 1981; Anati 1994).

Pictures of trees and structures made of wood also occur in great numbers on the picture stones on the island of Gotland in the Baltic. The picture stones dating to the period 400–1100 AD are a unique Gotlandic phenomenon; only three comparable stones have been found outside the island (Andrén 1993). On these stones – interpreted as gravestones and commemoration stones for dead people – a ship is generally depicted, but wagons and trees also occur (Lindqvist 1941–42; Nylén and Lamm 2003; Imer 2004).

Information from monuments and objects

Trees and wooden structures may also occur in the form of monuments. Anders Andrén has recently put forward an interpretation of the Scandinavian tricorns dating to 200–1050 AD, and according to him these should be understood as representations of trees, and the three arms coming out from the centre of the monument is a homology to the roots of the *Yggdrasil* tree known from Scandinavian mythology (Andrén 2004: 407–418).

A well-recognized monument in Scandinavian archaeology is ship settings dating from the Late Bronze Age to the Late Iron Age. These monuments are a translation of wooden ships into stone ships whereby some of these ships have constructional details that can only be understood against the background of wood craftsmanship (Capelle 1986; Berntsson 2005: 130–132).

Finally, it should be mentioned that representations of trees can also occur on objects, although this is little discussed (Andrén 2004: 418–421).

FROM TECHNICAL SOLUTIONS TO SOCIAL BIOGRAPHIES

When studying the literature cited above it becomes obvious that archaeologists have primarily been interested in technical solutions in connection with wooden structures. There is a broad literature on how ancient buildings were constructed, discussing topics such as the width of the aisle, the distance between the trestles and in what degree the roof rested on the aisle or the outer walls. When it comes to ships, there is a great interest in the development of the clinker technique, the introduction of oars and the sail. Wagons have been studied from an evolutionary perspective, with the introduction of the spoked wheel attracting interest.

This interest in the technical aspects of wooden structures has not been counterbalanced with an interest in how trees were used and treated by humans in the past. These questions are crucial for gaining a broader understanding not only of trees in prehistory but also of wood craftsmanship.

The study of vegetation has mainly been carried out by palaeoecologists, sometimes as part of archaeological projects. Palaeoecologists study trees as part of forest history from data such as pollen, macro and micro charcoal and sometimes insect remains. From their data it is possible to reconstruct the long-term vegetation history and also to discuss the question of human impact. However, the social significance of trees can rarely be constructed out of pollen or other materials found in peat and bogs.

To study the social significance of trees in prehistory we must study trees in a social setting – i.e. we must study the remains and representations of trees in archaeological contexts (Hastorf and Johannessen 1996). If we look at the archaeological record available, there seem to be three fields of special interest.

The first field is pictures of trees on Scandinavian rock carvings. These have often been treated in isolation, and the identification of tree on the carving has been a goal in itself. These pictures should be related to the humans and animals surrounding them in order to discuss the practical and ritual significance of trees in the Bronze Age.

The second field of interest is charcoal from archaeological sites, identified to species and ^{14}C-dated. Since these charcoal pieces have an age and a known position, it is possible to assign them to different categories, such as charred remains of trees from agricultural plots, graves and settlements, in order to gain insight into the different uses of trees in different contexts and how this changed through time.

The third field of interest is post-holes in buildings. By analysing the charred remains of wood in post-holes, information about the tree used in the construction may be obtained. Moreover, by comparing very similar houses built in sequences it might be possible to approach the question of timber reuse, since identical constructions might indicate the reuse of substantial parts of the timber structure.

Drawing on these three perspectives, it should be possible to discuss the practical, social and ritual significance of trees from an archaeological viewpoint. This book is an attempt to doing so. In the following prehistoric trees will be regarded as important landmarks in the landscape, loaded with meanings and with individual biographies. The individual character of trees and the meaning attached to trees will be studied in a social setting – i.e. the prehistoric landscape consisting of houses, fields, graves, trees, animals and humans.

An important notion running through this study is that trees have biographies. Trees are planted, kept alive, cut down and thereafter used in various ways. Branches and twigs might be used as fodder and leftovers not eaten by the animals as fuel (Regnell 2003a). The stem of the tree might be trimmed and used as timber in a building, and when the building is pulled down parts of the timbers might be used in the construction of another building (Welinder 1992). When a wagon is worn out, parts of it may be remodelled into other kinds of objects – a wheel axle might be turned into a shovel (Schovsbo 1987: 47).

As stated by Igor Kopytoff, any object can have many different biographies. A car, for example, has a technical biography summarized in the repair record, it has an economic biography that is its initial worth, its sale and resale price, the cost of several years of maintenance, and so on. The car also has a social biography focusing on its significance in the owner's family and its status as a social marker (Kopytoff 1986: 68).

All these biographies, whether they are technical, economic or social, may be culturally informed. What makes a biography cultural is not what it deals with but the perspective from which it is studied. A culturally informed biography of the practical use of trees will look upon trees as living organisms classified into culturally constituted categories endowed with a social significance played out both in rituals and in the everyday use of trees and timbers.

Therefore, in this study there is no definite limit set up between trees and timber. During a large part of the period studied trees were only slightly modified to become timbers in buildings. From the biographical perspective used here we are interested in the total life history of individual trees regardless of whether they are defined as living trees or dead timbers.

In the following the biography of trees and timber will be explored in three chapters.

Chapter 2 deals with the practical and ritual biography of trees as demonstrated by pictorial evidence from Scandinavian rock art and metal items. The significance of trees as sources of fodder by pollarding was a prerequisite for individual trees being loaded with symbolic meanings. The trees on the rock art have traditionally been regarded as spruce but their sometimes peculiar shapes are better explained if they are regarded them as culturally modified trees – i.e. shredded or laterally pruned trees. A new understanding of the pictured trees is thereby opened up – on the rock art we can identify trees in a landscape setting with humans and animals interacting with individual trees.

Chapter 3 deals with the landscape biography of trees. By using radiocarbon and species-identified charcoal from archaeological contexts in two different areas in the Småland Uplands, Sweden, the landscape biography from the perspective of trees will be highlighted. Depending upon the history of the landscape – whether it is a central area continuously inhabited for centuries or even millennia or whether it is an area that has been settled for a limited period of time – the use of trees will take on quite different forms. The character of landscapes is structured through people's practices. While doing things in the landscape, humans alter the relationship between grazing land and woodland and the composition and shape of trees which in turn affect the basis for their material culture.

Chapter 4 deals with the practical biography of trees and timber. Houses have often been studied as types with a definite shape. However, buildings are by nature composed of different materials that are used for specific constructional purposes. In prehistory various structural details were fastened together by rope and bast facilitated the reuse of older timbers. By analysing species-identified charcoal from an Early Iron Age house outside Malmö in Sweden, the use of various kinds of timbers for different purposes in the building will be detected. By discussing Bronze Age settlements from Holland and Sweden it will be argued that Bronze Age farms were stable units with houses built in a sequence on the same plot. By analysing the position of post-holes in these houses, sequences in the possible reuse of timber will be demonstrated.

Chapter 5 is a kind of summary where the different parts of previous discussions are brought together into a synthesis. The birthplace of Carl Linnaeus – the estate of Råshult – in southern Sweden is used as starting point for a discussion of the social significance of trees in the Bronze Age and Iron Age of northern Europe in a long-term perspective.

Chapter 2.
THE SIGNIFICANCE OF INDIVIDUAL TREES

The very close relationship between humans and animals has been used in many cultures to express and classify human-to-human relationships. According to Claude Lévi-Strauss, animals were good to think with because they could be ordered in contrasting sets and hierarchies that were analogous to the human world. This relationship between humans and animals is expressed, for example, in the totem identification of humans with animals among certain groups of native North Americans, and also among native people in Australia (Lévi-Strauss 1966; Morphy 1991; Bloch 1998).

Like animals, trees have certain characteristics that make them useful as metaphors in the endeavour of humankind to understand the relationship between humans and the surrounding world. Trees are alive but their lives are very different from that of humans and animals. They lack mobility, they do not consume food, and their rhythm is based on the year rather than the day. Through the fact they are living and growing they share characteristics with humans, but the differences between humans and trees are greater than those between humans and animals.

This ambiguous status of the tree – close to man and alive but also very different and almost non-living – is often explored in creation myths where trees act as mediators between the non-living world and the living world. The creation of women and men out of trees is a well-known theme; this is facilitated by the fact that trees lack an obvious sex and therefore they are easily associated with both women and men. Trees are often believed to be hermaphroditic and offer an excellent way to symbolize the reproductive couple – i.e. the genderless potential of self-regeneration that comprises both female and male life principles (Rival 1998a: 10).

The identification of humans with trees is further facilitated by the close resemblance in shape between trees and humans, as expressed in the associations between, for example, sap and blood, leaves and hair, limbs and arms, bark and skin and the stem and the human body. These identifications arise from a very close relationship between humans and trees, where trees play an important role as sources of food and materials (Rival 1998a: 10).

In pre-modern European societies people were engaged with trees in much more sophisticated way than today. Trees were part of an annual cycle in which leaves, twigs and branches were collected during summer and early autumn and stored as winter fodder for the animals. As people spent days and weeks in the field sowing and ploughing, they also spent days and weeks climbing around in trees cutting twigs and branches (Tillhagen 1995; Emanuelsson 1996; Kirkby and Watkins 1998; Rackham 2003; Slotte 2000; Watkins 1998).

Through practice people gained a knowledge of trees that was very different from the view of modern man. Working with and observing trees was part of everyday life. As stated above, trees also had certain characteristics that made them good to think with since their qualities made them useful as mediators between earth and heaven and as metaphors of human life. Real trees were therefore represented in other media as pictures on rock carvings, as monuments of stone and earth or as artefacts with the shape of trees (Andrén 2004).

Even though symbolic representations of trees dominate quantitatively, in some cases we can also study man's relationship to real trees. Two British studies discussing man's relationship to tree throws in the Early Neolithic and the ritual use of a real tree from the transition between the Neolithic and the Bronze Age, are revealing.

THE RITUAL SIGNIFICANCE OF TREES

Tree throws are now and then recorded on archaeological sites; when these are found on Mesolithic or Neolithic sites there are sometimes difficulties in distinguishing between the remains of tree throws and huts. Instead of making a clear distinction between natural tree throws and dwellings made by man there are other views stressing that three throws might be the result of man pulling down trees with ropes (Brennand and Taylor 2003: 62), tree throws being used as foundations for dwellings (Evans *et al.* 1999: 249) or incorporated in Neolithic cursuses (Buckley *et al.* 2001: 152–153) and artefacts being deliberately put into tree throws, as argued by Evans *et al.* (1999).

The latter perspective is put forward as an explanation for the extensive findings of Early Neolithic material in tree throws excavated in Oxfordshire by the Oxford Archaeological Unit. At Hinxton a tree throw measuring 4.30 x 3.20 metres was excavated, indicating that a mature tree with a trunk about 0.50–1.00 metres and a root ball of 3–5 metres across, could have stood on the spot (Evans *et al.* 1999: 242–244).

Yet another two tree throws were excavated at Barleycroft Paddocks. In all the three tree throws a larger amount of pottery including both fine ware and storage pots, flint artefacts and flint debris were found dating to a very early part of the fourth millennium BC. Fieldwalking and excavations revealed no finds in the subsoil and in the features close to the tree throws, but at some distance from the tree throws at Barleycroft Paddocks pits with Neolithic material came to light; however, these did not show the same scale of artefact deposition (Evans *et al.* 1999: 242–248).

The location of the material indicates that it was deposited after the tree had fallen and not when it was standing (Evans *et al.* 1999: 248). We thereby get a glimpse of how the fall of a mature tree created a place of special significance where the deposition of artefacts helped to reinforce the relationship between man and individual trees. In fact, there might be a practical background to the depositions in the tree throws. The native people of British Columbia used fallen trees as resources for various purposes:

> It was also a common practice to take planks from windfall trees, which is not surprising, considering the work that would have been involved in felling an entire tree. This practice conserved, on the one hand, a great deal of time and energy and, on the other, the life of a tree (Stryd and Feddema 1998: 9).

From this perspective the depositions of artefacts in tree throws might be seen as a mark of gratitude, as the fallen tree may have represented a substantial value. There is yet another example from Britain that enables us to shed further light on the special treatment of fallen trees.

Because of erosion and tidal movement, a timber circle eroded out of the beach at Home-next-the-Sea close to Norfolk in 1998. An excavation revealed a timber circle with a diameter of 6.78 metres made of timbers, 0.15–0.40 metres in diameter; most of them were half-split timbers but round-wood timber also occurred. A forked branch put upside down probably served as an entrance to the structure. A combination of ^{14}C dates and dendrochronology dates the construction to 2049 BC (Pryor 2002; Brennand and Taylor 2003; Watson 2005).

In the centre, or slightly south-west of the centre, stood the bottom of an inverted oak with roots measuring 1.20 metres in diameter. The survival part of the tree shows no traces of being felled by axes, and therefore the tree was either blown over or felled by rocking. In this respect the central tree differs from the trees making up the circle, which show clear signs of having been felled by axes.

This inverted tree might therefore have had a special significance already as a standing tree which necessitated special treatment – i.e. intentional toppling – when being felled. Alternatively it was the felling by storm that lent it a special meaning; or it was the combination of the two. No matter which is the case, the inverted oak tree from Home-next-the-Sea is a very good example of how individual trees gained specific meanings that gave them a crucial position in rituals and ceremonies.

In this sense the use of trees in rituals echoes how other elements in the agricultural practice were used in ritual settings. Richard Bradley has drawn our attention to how storey houses and granaries in prehistoric Europe took on a ritual significance that went far beyond their domestic importance. Granaries are known from archaeological excavations and historical sources, and differ from ordinary buildings in both function and shape. As they were storehouses it was important that the stock of food was kept dry and away from animals. They were therefore given a raised shape with the floor on stilts and the door up off the ground (Bradley 2005: 98–100).

Buildings with these qualities were depicted on rock art in Valcamonica in northernmost Italy and they were transformed into ceramic models where the door does not open at ground level but is situated at a level above ground. These ceramic models, often called house urns, are distributed in northern Germany, Scandinavia, the Baltic and adjoining areas (Sabatini 2007). They had a clear religious significance as they were used as containers for human cremations. They occur in gravefields dating to the 8th and 9th centuries BC (Bradley 2005: 98–100). This example reveals how certain elements in agriculture, such as the harvesting and storage of seed, had qualities that made them useful as metaphors in relation to the life and death of humans.

In recent years there has been a trend in archaeology to underline the importance of ritual action. Fundamental for ritual action is that it is repetitive, formal and regular; that it is part of a strategy; and that it is performed by a group of people that also have other relations and interests in common beyond the ritual. The shared experience of action rather than a common shared belief is at the centre of the ritual (Bell 1992; Humphrey and Laidlaw 1994).

This approach puts the stress on the ritualization of various kinds of actions and thereby provides a link between the actions of everyday life and formal rituals. As expressed by Richard Bradley:

> Once it is accepted that ritual is a kind of practice – a performance which is defined by its own conventions – it becomes easier to understand how it can occur in so many settings and why it may be attached to so

many different concerns. Once we reject the idea that the only function of ritual is to communicate ritual beliefs, it becomes unnecessary to separate this kind of activity from the patterns of daily life. In fact, rituals extend from the local, informal and ephemeral to the public and highly organised, and their social contexts vary accordingly (Bradley 2005: 33).

In expanding our understanding of the relationship between humans and trees – in both practical and ritual terms – we shall now turn to pictures of trees. Depictions of trees in Scandinavian rock art and on metal items are useful – but very seldom considered – sources for understanding prehistoric man's relationship to trees in everyday life, in rituals and in mythology.

THE PICTORIAL EVIDENCE – PROMISES AND OBSTACLES

Scandinavia, together with the Alps and Atlantic Europe, is one of those regions in Europe that stands out as an area with many rock-art sites (Malmer 1981; Anati 1994; Bradley 1997). In contrast to Atlantic Europe but like the Alps, the Scandinavian motifs are often highly figurative, enabling us to recognize features like people, animals, ships, weapons, wagons and ards (Goldhahn 2010).

However, the density and character of motifs differs between various countries and regions within Scandinavia. Traditionally two different rock-art styles are identified: one in northern Scandinavia focusing on animals and one in southern Scandinavia where the ship plays a prominent role. Often these two traditions are also connected to different ways of living; hunter-gatherer populations are supposed to have fabricated the carvings in the north while a population dependent upon agriculture is supposed to have made the carvings in southern Scandinavia (cf. Malmer 1981: 3).

Recently this view has been questioned and it has been argued that in many regions in southern Scandinavian peoples primarily identified themselves as fishermen and seafarers and not as agriculturalists. The argument is that the bulk of rock art occurs along the coasts in the Baltic and the North Sea; and

when the Bronze Age shoreline can be reconstructed it turns out that most of the rock art was executed adjacent to the shore line; and even sometimes on almost vertical cliffs that only could be reached by boat. Moreover, in numbers the ship totally dominates the panels (Ling 2008; Skoglund in press).

In contrast to this situation there is yet another situation occurring in present-day Denmark and southernmost Sweden. Here rock art was used in a different way: the amount of rock art is much smaller, there is less emphasis on the ship, and the motifs are primarily located on smaller stones that were often incorporated in graves or on certain occasions cult houses (Glob 1969; Kaul 1987; Skoglund in press).

There is another difference between the Baltic and the North Sea on the one hand and Denmark and southernmost Sweden on the other hand. In the latter area figurative motifs on metal items are more common than in the rest of Scandinavia; some of these items, like the metal razors, are comparatively often depicted with different kinds of figures and motifs that also occur in rock art (Kaul 1998a).

The conclusion is that in Denmark and southernmost Sweden figures were primarily depicted on personal metal items or on stones that were incorporated in monuments while in the Baltic and the North Sea figures were commonly displayed on panels in the coastal landscape. This suggests that the motifs were part of different rituals and social settings – i.e. a more individual oriented agro-pastoral society in Denmark and southernmost Sweden in contrary to a more maritime and public milieu further to the north (Skoglund 2009).

The depictions of trees proper, where individual branches can be identified, are low in number, amounting to roughly 30, but only two of these are found in Denmark. The depictions of trees follow the general pattern discussed above. The two Danish examples are found on portable stones (Glob 1969: 96–103, nos. 14, 91); while the rest of the trees from other parts of Scandinavia are depicted on panels in the open air (Nordéen 1925: 43, 47–48, 56, 95, 191, Pl. XXIV, Pl. XXVIII, Pl. CXXV; Almgren 1927: 12, 14–15, 17, 102; Marstrander 1963: 371, Pl. 31; Fredsjö et al. 1975: 43; Burenhult 1980: 9: 67: Fredsjö et al. 1981: 223; Högberg 1996: 36–37; Andersson 1997: 5, 9, 11, 13, Pl. I, Pl. V, Pl. VII, Pl. XI, Pl. XIII; Hygen and Bengtsson 1999: 161; Bengtsson and Olsson 2000: 35–36; Skoglund 2006: 54–55).

Most trees occur in two concentrations: there is one concentration of trees in Bohuslän on the North Sea coast and one in Östergötland in eastern central Sweden. About 10 trees are known in each of these areas. Moreover, in northern Bohuslän, on the west coast of Sweden, with the largest concentration of rock art in Scandinavia, we also find the most elaborate tree depictions. In this area individual trees are depicted but we can identify trees in a landscape context and activities carried out in connection with trees.

Given that many of the motifs on the Scandinavian rock-art sites can be identified with objects known from archaeology or anthropology, there is a long tradition in the research history of identifying and interpreting individual motifs in terms of existing objects, as if there was a single and fixed meaning to be attached to the figures. This approach has been criticized and it has been argued that rock-art motifs are ambiguous, with multiple layers of meanings attached to them (Goldhahn 2005: 63–156, 2006: 89–90).

This perspective gets support from the study of the iconography on the contemporary metal objects where abstract decoration sometimes take a more definite shape and can be interpreted as specific symbols known from other contexts. There is also the situation when animals are composed of characteristics from two different kinds of species (Kaul 1998a). In these cases there is a clear ambiguity, with the interpretation of a motif depending upon the situation and the observer. This is logical if we regard motifs on rock-art panels and metal items as bearers of complicated information with different layers of meaning attached to them. This might reflect a situation where one kind of knowledge, public for everyone's use, was complemented by specific knowledge guarded by ritual specialists (Skoglund 2009).

There is yet another perspective that is important. Even though motifs were composed of details from different contexts and were made highly stylized, the motifs were based on actual observations of the surrounding environment. Perhaps the best way of illustrating this is the relationship between ards depicted on rock panels and preserved wooden ards found in bogs. A comparison of these two sources of information makes it obvious that the people who depicted the ards on the rock used their knowledge of existing wooden ards to produce an accurate picture (Glob 1951; Bradley 2005; Skoglund 2008). Similar kinds of conclusions can be drawn when comparing depicted ships (Kaul 2003) and wagons (Nilsson 2005) with ships and wagons known from anthropological and historical sources.

Therefore, when looking at rock art or figures engraved on metal items we do not look at pictures that try to resemble reality. We are looking at a ritual expression that is composed of observations from reality, but these pieces of information are used to create figures that are logical according to a specific ritual, and not according to how the real world is constructed. However, to gain an understanding of these motifs – based on an archaeological and contextual approach – we must begin by trying to understand what kind of observations form the surrounding world that lay behind the motifs.

When it comes to rock-art images representing trees, very little has been done in trying to gain an understanding of how these trees can be related to the Bronze Age cultural landscape or how they are related to insights gained from anthropology. Instead the trees have been regarded as exotic objects interpreted as evergreen spruces or yews that lacked any relationship to everyday life. In the following this perspective will be questioned.

Before proceeding, a brief note on rock-art chronology is needed. The tradition of producing rock art in southern Scandinavia has a long tradition going back at least to the transition between the Neolithic and the Bronze Age, c. 1700 BC, and ending around the birth of Christ. Recent research based on comparisons between motifs on rock art and motifs on metal indicates that individual panels consist of motifs of different ages (Ling 2008).

It is therefore not possible to draw conclusions from motifs occurring together if they cannot be regarded as meaningfully connected by composition. A composition could, for example, be if several people are involved in the same procession and share characteristic attributes or if different people in the same panel carry out activities that are logically related to each other.

SPRUCE OR DECIDUOUS TREES?

In 1927 the Swedish archaeologist Oscar Almgren wrote a book that has influenced the interpretations of Scandinavian rock art ever since. Almgren argued that the rock art should be viewed from a ritual perspective and that the scenes depicted on the rocks reflect rituals carried out in reality. To emphasize the point he made extensive comparisons with material from both Europe and the Near East. For example, he demonstrated that the Scandinavian rock-art scenes with a man carrying a ship could be understood against the background of European carnivals and ancient Egyptian rituals, where those ships were carried around by people in rituals (Almgren 1927).

Almgren has not only influenced the general idea of how Scandinavian rock art should be viewed, but he has also had a great impact on the understanding of specific details on the rock art such as the tree. In his book Almgren stated that most of the trees depicted on the rock art represented spruce. This conclusion was based on the similarities in shape between the rock-art images representing trees and the spruce (Almgren 1927: 17).

Since 1927 the number of depicted trees has increased and today there is a greater variation between the individual trees than Almgren was aware of. Despite this, Almgren's suggestion that the trees represented spruces has not been questioned. The general idea ever since Almgren's work is that the rock-art trees are conifers, i.e. spruces or yews (cf. Hygen and Bengtsson 1999; Andrén 2004: 404). However, there seem to be arguments in favour of an alternative view recognizing these trees as culturally modified – i.e. shredded deciduous – trees.

The first argument is the very variety of shapes among the rock-art trees (Figure 4). Among these are trees that resemble spruces, but there are also trees that partly lack branches, or where the branches look different from what could be expected on a spruce. This variation can only be explained if one accepts these trees as shredded. When depicting shredded trees, it is logical that in some cases the trees are shown with branches from the bottom to the top, in other cases the trees are shown with only a few branches.

The second argument concerns the relationship between trees and people (Figure 5). In some of the trees there is a human depicted at the top of the tree. It is of course possible to climb a conifer but it does not make sense in practical terms. In contrast, you have to climb a shredded tree in order to utilize the branches. The occurrence of people at the top of the trees therefore speaks in favour of these trees being shredded.

The third argument concerns the migration of spruce into Scandinavia and the absence – or very limited occurrence

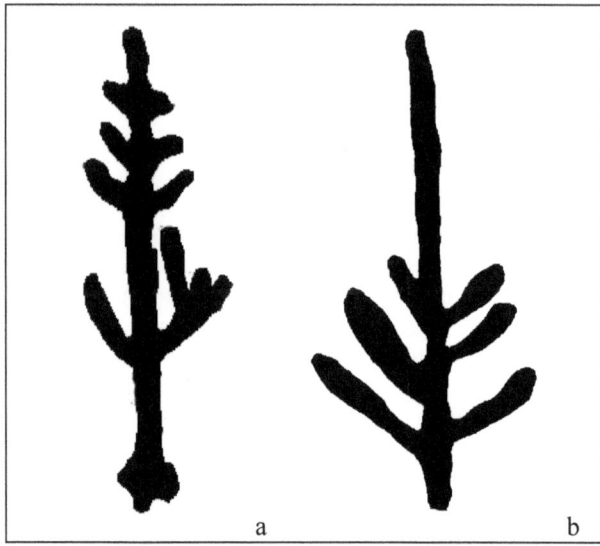

Figure 4. a) A rock-art tree from Kyrkorud, Tanum, b) a rock-art tree from Ör, Småland. Information from Bengtsson and Olsson 2000: 45–46 and Skoglund 2006: 54

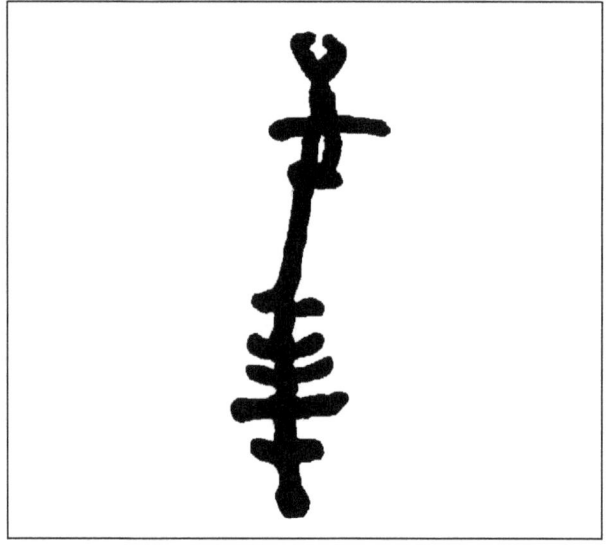

Figure 5. Human at the top of a tree. Tanum; Bohuslän, RAÄ 66. (The outcrop of 1,000 cup-marks)

– of spruce in the Bronze Age landscape. In contrast to all other trees in the Scandinavian peninsula, spruce migrated from the north to the south. From pollen analysis we know that spruce joined the Scandinavian flora late. It was established in the Oslo and Stockholm area at a larger stand-scale around the birth of Christ and it moved southwards with a speed of 150–200 metres a year. It never migrated naturally into southernmost Sweden and Denmark; the occurrence of spruce in these areas is the result of plantations in the 19th and 20th centuries (Ohlson and Tryterud 1999; Giesecke 2004; Giesecke and Bennet 2004; Lindbladh 2004).

It should be noted that we know from pollen analysis that we cannot rule out the possibility of individual spruces in south Scandinavia during the Bronze Age; but if they existed spruce was a very rare tree in the landscape.

Accepting that the pictures represent shredded deciduous trees opens new perspectives on the relationship between man and trees in the Bronze Age.

THE PRUNING OF TREES

Pollarding, coppicing and pruning of trees are well known practices in traditional European societies. Today the collection of leaf fodder mainly survives in rural areas in southern and eastern Europe; but before the industrial revolution in the 19th century coppicing trees and collecting leaf was, together with haymaking, the standard way of feeding the animals during the winter (Tillhagen 1995; Emanuelsson 1996; Kirkby and Watkins 1998; Watkins 1998; Slotte 2000; Rackham 2003)

The leaves could be reaped off the trees by hand or more commonly twigs and branches were cut with a knife, saw or axe. When the cutting was over the branches were bunched together in bundles. Either they were hung individually on stakes, or at standing trees, to dry, or many bundles were stacked together in a leaf-stack. To stabilize the leaf-stacks – which could take on impressive dimensions – a skeleton of poles was sometimes built to keep the bundles in place. These leaf-stacks could either be located where the pruning took place, or alternatively, the bundles were transported to the farm where the leaf-stack was built close to the stable (Halstead *et al*. 1998; Slotte 2000).

In times of dearth – or when there was a limited range of trees available – all kinds of trees could be used for coppicing. However, there seems to be a general consensus that leaves of ash, elm and lime (linden) were preferred as animal fodder. Other trees mentioned in this context are birch, rowan and less often oak (Borgegård 1996; Slotte 2000).

Basically there seem to be three ways of manipulating trees to facilitate the collection of leaf-fodder (Figure 6) (Rackham 2003; Bergendorff and Emanuelsson 1996).

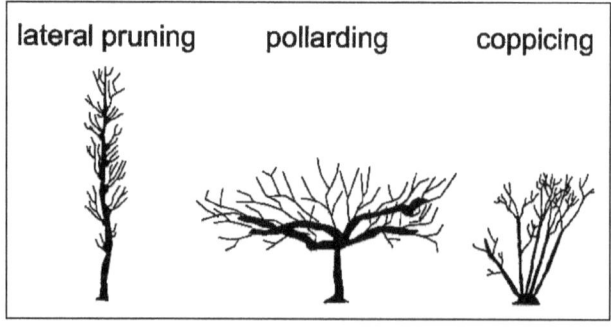

Figure 6. Sketch explaining different ways of pruning trees

The tree could be pruned laterally. In this case the branches were cut from the stem, making the tree look pole-like. These trees were allowed to reach their natural

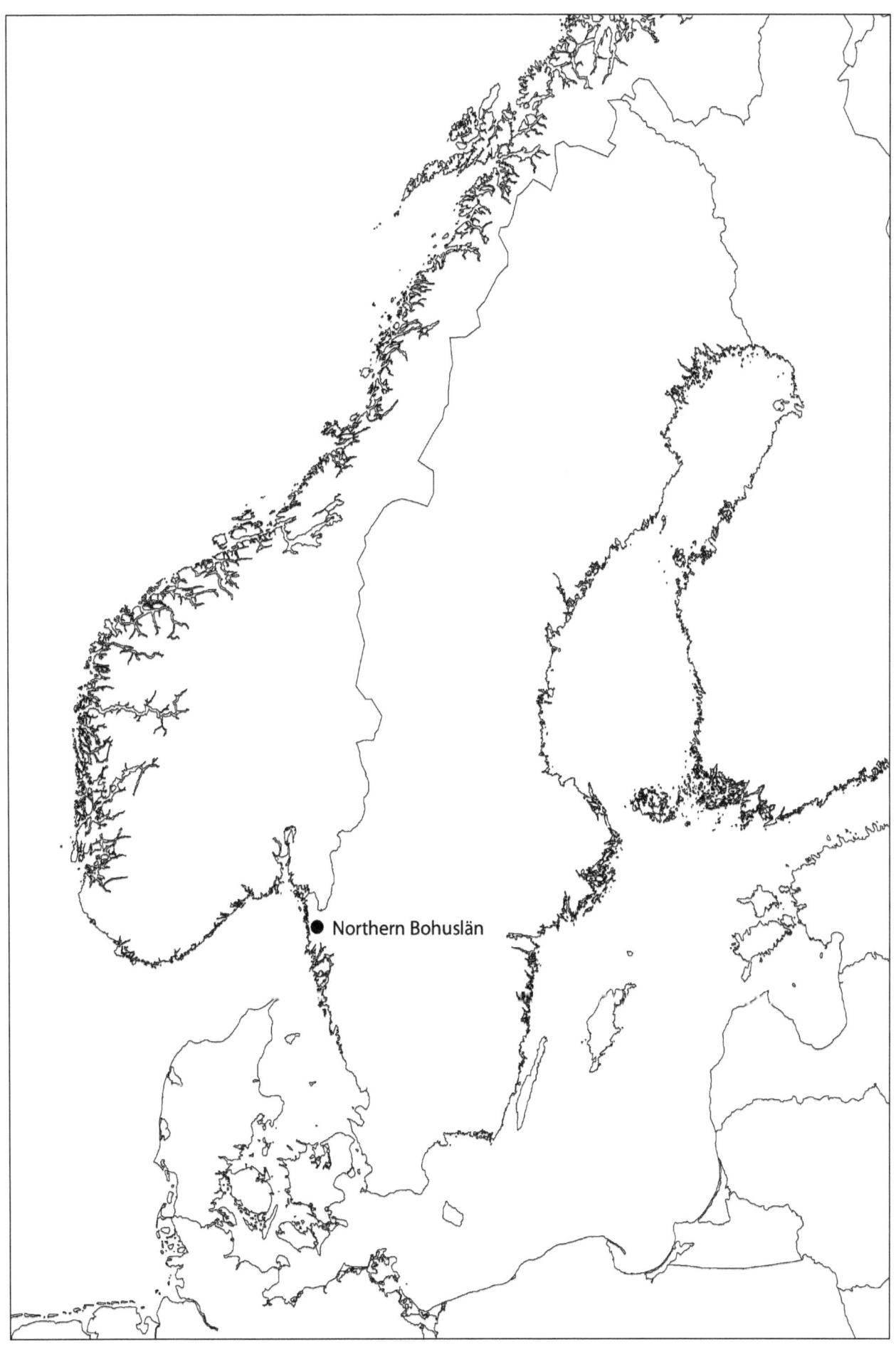

Figure 7. Map showing the location of northern Bohuslän in Sweden

height, at which it involved some danger to collect the foliage. The top of the tree was reached by climbing on the branches, or the small stumps left after the branches were cut away.

The tree could be pollarded. In this case the top of the branches were cut back at a height of 2–3 metres, i.e. above reach of grazing animals. The tree took on the shape of candelabra. The branches got closer to ground and could be reached by using a ladder.

The tree could be coppiced. In this case the tree was cut off close to the ground which triggered the production of branches from the stump. This way of manipulating a tree is less often used when collecting leaves only, but is an excellent way of getting many straight branches of similar sizes that could be used for construction.

These practices were widespread in Europe before industrialization and can be studied in a pre-modern context in paintings illustrating agricultural work (Hæggström 1996). From northern Germany and Scandinavia we know of specialized tools for collecting leaves dating to the time span 200 BC–200 AD (Penack 1993). These are iron knives with a crescent-like shape, making them suitable for cutting twigs and branches from trees. Before this leaf collection was most probably done with axes such as socket bronze axes and before that with flint axes. Knives and sickles of flint and bronze were probably too fragile to be used for cutting branches.

The use of axes to cut down branches supports the idea of a widespread use of leaf fodder in Neolithic and Bronze Age Europe in comparison to haymaking. Efficient use of grass for fodder requires a scythe. The scythe was used by the Romans and it was known in Central Europe at the birth of Christ. A short scythe that later developed into a longer and more efficient tool came into use in Scandinavia around 200 AD (Myrdal 1984: 87). Because efficient haymaking was first made possible with the invention of the scythe, the importance of leaf collection was probable greater before the Roman period.

Investigations of well-preserved Alpine lake-shore settlements dating to the Neolithic and the Bronze Age shed further light on the collection of leaf fodder in prehistory. From the analysis of twigs and faeces found in cultural layers at these settlements we know that goats and/or sheep were fed with leaves and twigs during the winter and early spring when there was a snow cover (Haas *et al.* 1998; Karg 1998; Akeret 1999). Similar results are reported from other parts of Europe in those rare cases when there are comparable preservation conditions (Göransson 1995).

Occasionally pictures on Egyptian, Greece and Roman frescos and vases inform us of trees that look as if they have been coppiced (Hæggström 1996). A general problem with these pictures is that it is often difficult to clearly state that the trees are coppiced since they are highly stylized. Moreover, they do not inform us of the leaf collection process. From this perspective the Scandinavian rock-art motifs depicting trees and humans are important since they give us insight into the interaction of trees and humans: on these motifs we can see humans climbing the trees, cutting down branches, building leaf-stacks and feeding their animals.

In the following this will be elaborated by discussing three complex scenes related to leaf collection from the northern part of the province of Bohuslän in Sweden (Figure 7).

RITUALS RELATED TO TREES IN THE LANDSCAPE

Runohällen – cutting branches and constructing a leaf-stack

Runohällen – the runic panel – is located at Ryk in Bohuslän some 30 metres from the Gerum river. The panel is rather large, measuring 9 x 6 metres, and parts of it are inclined at a considerable angle. On the panel are ships, humans, animals, foot soles, ring crosses and circles depicted together with cup-marks (Baltzer 1885: pl. 39–40; Bengtsson and Olsson 2000: 36; Fredell 2002, 2003: 163–167, pl. XIII)

By measuring the altitude of the panel and comparing the data with what is known of the shore displacement for this part of Scandinavia it is possible to gain some idea of the dating of the panel. This work carried out by Johan Ling reveals that the cliff rose out of sea during the period 1700–1300 BC (Figure 8). Another way of dating the panel is through the ship images. The chronology of the Scandinavian ship images suggests that the first ships on the cliff were made during Montelius period II, 1500–1300 BC, but that new motifs were added all throughout the Bronze Age (Ling 2008: 87ff, 136ff).

A famous and much discussed motif is a scene in the centre of the panel referred to as "the maypole". Almgren interpreted this image as a tree being drawn by people holding ropes in their hands. He compared the scene to a tradition known from 19th century Sweden where trees were drawn on sledges and placed close to farms where weddings were celebrated (Almgren 1927: 104f). Another interpretation is that we see a pole or a tree where ropes are tied to the top and fastened on the ground by the people bending their backs (Fredell 2002: 253).

This motif has often been interpreted in isolation, but as Åsa Fredell has pointed out, it is connected to three other scenes by the repeated occurrence of people wearing masks with horns or by objects splitting into three branches. This notion is important because it indicates that we should see this scene as a process involving several elements rather than as a static element. Fredell interprets the motifs as a process going from bottom to top and ending with an offering scene where the three-branched objects are submerged in the sea (Fredell 2002: 252–253).

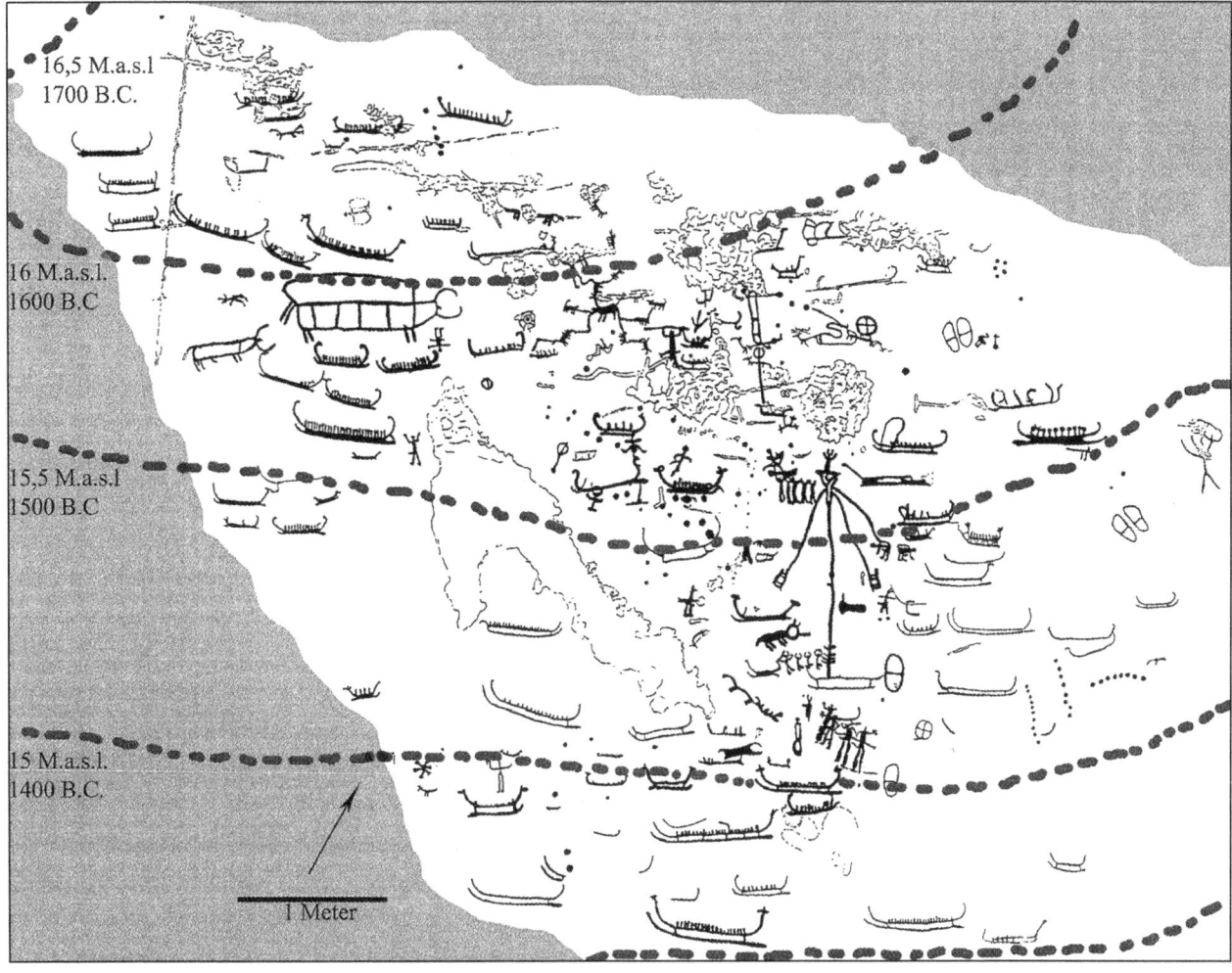

Figure 8. Runohällen with measured terrain curves indicating when the cliff rose out of the sea.
From Ling 2008: 91

Generally the interpretations of the presumed maypole and the motifs surrounding it are characterized by a lack of will to discuss these kinds of rituals against the background of man's everyday interaction with trees. Looking at these scenes from the perspective of pollarding will add substance to the interpretation of the presumed maypole (Figure 9).

Below the presumed maypole we find a tree. From the stem of the tree there are branches growing, making this tree look like a tree pruned laterally. The tree is connected to the presumed maypole with a line that might symbolize the tree as shredded – i.e. when the tree takes on a pole-like appearance – but this is an uncertain assumption. To the left of the tree are four men with horns. People wearing horns are well known from Scandinavian rock art and the horns, or more probably a helmet with horns, signify their status as ritual specialists (Goldhahn 2007).

Above and to the left of the presumed maypole we find a tree and at the top of the tree a human with raised arms. The interpretation of this as a tree with a human is based on the similarity of this motif to trees where the branches are clearly visible. A very clear picture of this situation is found at Aspeberget (see the discussion below). Here we find a tree where branches are lacking in the upper part of the tree (as if they were cut away), but clearly visible in the lower part of the tree. At the top of the tree is a human with arms stretched out and adorned with horns on the head. The situation at Runohällen very much resembles the situation at *The outcrop of 1000 cup-marks* (Figure 5 and discussion below), but at Runohällen all the branches are cut away, and instead we find branches lying on the ground symbolized by the three-branched objects.

Below the opening scene we find these objects again – now in the hands of four people walking in a row. This makes sense if we presume that bunches of branches are being transported from the place where they were cut down, to a place where they could be dried and stored for consumption during the winter.

Finally there is the presumed maypole. The idea that the lines represent ropes connected to a pole does not seem convincing. Instead we get an impression of four poles joined together and supporting a small floor attached to the top of the poles. Following the line of the earlier interpretations, we may presume that it is a leaf-stack constructed of poles to stabilize bunches (Figure 10).

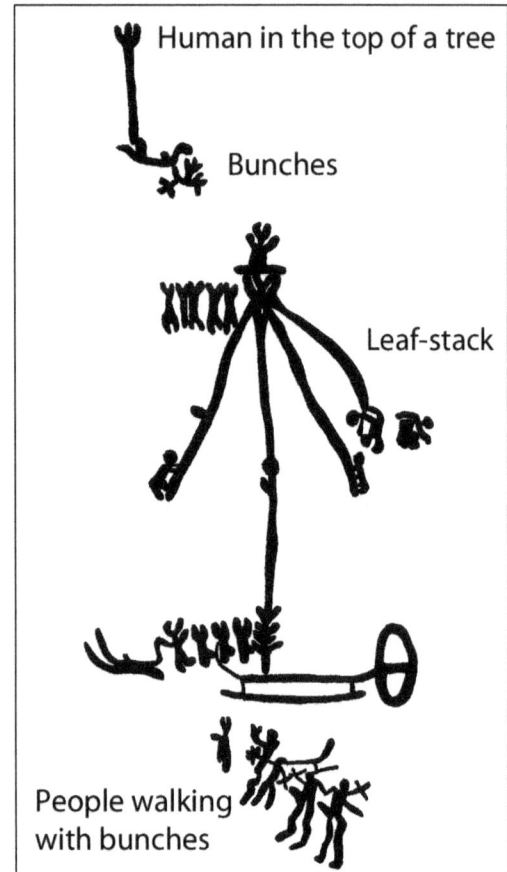

Figure 9. Detail of Runohällen. Information from Fredell 2002: 252

Figure 10. Leaf-stack on a photo from 1931 from Slätthög Parish, Småland Uplands. Photo provided by Folklivsarkivet, Lund

Even though we cannot see the bunches themselves there are a lot of activities going on in this scene – a person is bending over and stabilizing a pole in the ground, another person is bending, as if pulling something up from the ground, and finally two men seem to be starting to climb the poles as if they are moving upwards the leaf-stack. All these actions can easily be associated with the construction of a leaf-stack where a pole is being set in the ground and bunches of branches are being transported from the ground to the stack. Finally, it may be noted that a similar scene with two people climbing a pole is to found at Härkeberga in Upland (Coles 2000: 67–68).

Aspeberget – feeding the animals

Aspeberget, a hill measuring 250 x 200 metres, is situated adjacent to the Tanum plain in Bohuslän. The hill is famous because of its many rock-art panels with a variety of motifs situated on different parts of it. In this context we are interested in one of the panels on the top of the hill, well known because of its five very clearly depicted bulls (Baltzer 1881 pl. 23–24; Fredell 2003: 130–137, pl. 7).

This part of the hill is located too far up to allow any chronological assumptions to be drawn from the shore displacement. Judging by the ship chronology, most of the motifs on this part of the panel was made during Montelius period III–V, 1300–700 BC (Fredell 2003: 133).

At Aspeberget we again meet the three-branched objects (Figure 11). Here we find six of them connected to a circular plate often interpreted as a sun disc with people connected to it. Bearing Runohällen in mind and the identification of the three-branched objects as branches another possibility has to be considered – could this motif likewise be interpreted as activities carried out in connection with leaf fodder?

Accepting the idea of the three-branched objects as branches the logical conclusion is that here too we are dealing with a leaf-stack but this time the stack is reproduced from two directions – from the top as well from the side. The perimeter of the stack seen from above is symbolized by the circular plate, while six branches and four individuals are depicted from the side. The two people to the left have ponytails which are conventionally interpreted as a sign of female gender. The four people are apparently working with the leaf-stack; whether they are constructing it, or removing braches to feed the animals to the right of the stack, cannot be determined.

To the left we see three animals with long necks but without horns, allowing us to presume that they are horses. As pointed out by Åsa Fredell, the largest horse is pregnant (Fredell 2003: 133). Directly below this horse we see a person with its left arm raised towards the stomach of the horse as if it is milking the mare. A pregnant horse giving milk to humans is a kind of animal that would need extra nourishment, which strengthens the relationships between the group of three horses to the left and the presumed leaf-stack to the right.

A third scene related to this can be found to the left, where the group of horses has embarked a ship. The mare is no longer pregnant and another small animal, presumably the foal, has now joined the group. The four strokes on the ship symbolize people and indicate that the people previously nursing and milking the horses have also embarked on the ship together with the animals.

Figure 11. Part of the panel at Aspeberget

As a counterpart to this scene we find the group of five bulls, three of which are clearly oriented towards the largest ship on this panel. There is also a man tilling the soil with a bow ard. The ship is interesting since it has many connotations of animals. The horse head at the prow wears a mask with bulls, making this creature into something in between a horse and a bull. Another horn is seen on the lower part of the prow. In the stern there is another horse head.

The crew is depicted in a stylized way. However, three people differ from the rest of the group, a person in the middle that might be an animal and the two persons at the prow and the stern. These two people are wearing something in their hands that they are holding out towards the horses' heads. The object to the left is too stylized to enable any conclusion about its function, but to the right we recognize the three folded objects that were previously identified as a branch. What we see is a man feeding a horse, which means that this ship is partly made up of living animals.

The outcrop of 1000 cup-marks – leaf-stacks in the landscape

This panel is 7 x 4.5 metres large and is dominated by cup-marks (Högberg 1995: 36–37). According to Johan Ling, the motifs were created over a long period of time covering most of the Bronze Age and probably also the Pre-Roman Iron Age. The two larger ships at the centre of the panel are probably from the Early Bronze Age. The one to the left has an animal head which, according to comparative ship chronology, indicates Montelius period III, 1300–1100 BC (Ling 2008: 130).

The overall impression of this panel is that we are dealing with a landscape scene comprising both land and water (Figure 12). The amount of land-living animals and ships is quite equal, numbering 36 and 31 respectively. On this panel the cup-marks are used to make figures and compositions. A majority of them are found at central of the panel, making up a distinctive cluster. Among the cup-mark cluster are two animals. Four ships are attached to the cluster of cup-marks but no ship is found inside the cluster. It therefore seems possible that the cup-marks in this context represent land and the four ships are arriving at or departing from land. If the cup-marks represent land there is the possibility that the individual cup-marks represent trees seen from above, but this is merely a guess.

To the right at the panel there is a very nice tree depicted with a man at the top wearing a mask with horns (Figure 5, 12). The tree is pruned laterally; at the top the branches have been cut away but lower down they are still attached to the tree. The human is wearing a helmet with horns in the same way as the people in the panel at Runohällen.

Perhaps there are yet another similarity between this panel and Runohällen. At Runohällen there was a leaf-stack constructed of the branches cut away from the tree. In this context the leaf-stack could have been given a more stylized appearance – i.e. the circular arrangement of cup-marks. On the panel there are several such arrangements, but there are two distinctive groups: there is a group of two to the right surrounded by animals and there is a group of five to the left also surrounded by animals. Especially to the left we can see how each of the

Figure 12. The outcrop of 1000 cup-marks. From Högberg 1995: 37

three animals is interacting with a circular arrangement: their heads are oriented towards a circle or stuck into a circle.

Accepting that these are leaf-stacks, the five circular arrangements of cup-marks to the left would represent the situation in winter with a snow cover and animals being fed from leaf-stacks, while the motifs to the right, where the animals seem to be ignoring the leaf-stacks, would represent the situation in summer when bunches of branches are stored in leaf-stacks for the winter. In support of this there is a larger animal to the right, probably a bull with his head close to the earth, as if he is eating grass from the ground.

THE CONDUCTOR OF TREE RITUALS

Hitherto the discussion has focused on the practical aspects of leaf collection. From the analysis of the motifs on the rock art, however, it clear that we are not witnessing a true-to-life description of man's interactions with trees, but rather rituals where the collection of leaves was a substantial element. If we focus on the people participating in these rituals it is possible to put the scenes discussed above in a wider social and ritual context.

A common denominator for many of the people depicted at the top of the trees or in associations with trees is that they are wearing horns. This was true for most of the people at *Runohällen* and for the person at the top of the tree at *The outcrop of 1000 cup-marks*. People wearing masks with horns and blowing lurs also occur in close connection to a ship with two trees aboard at Kalleby, also in northern Bohuslän (Almgren 1927: 14).

From north Zealand in Denmark there is a unique find of masks made of bronze with horns, probably dating to 1100–700 BC (Thrane 1975: 67). People wearing masks with horns are occasionally found on bronze miniatures, and people with horns are represented on Scandinavian rock art (Goldhahn 2007: 331–341).

Various researchers have identified a set of objects that signify a male ritual specialist – among these are the mask with horns, accompanied by masks with bird's heads, cult axes, extraordinary costumes including a tongue-shaped apron, and gestures like the hand with spread fingers (Kaul 1998a: 16–35; Kristiansen and Larsson 2005; Goldhahn 2007: 331–341).

It is these male ritual specialists we find standing at the top of trees in some of the rock carvings in Bohuslän. The realistic depictions from Runohällen and Aspeberget of people climbing trees, cutting branches, building leaf-stacks and feeding animals with fodder goes back to an everyday experience – but what is depicted is the ritualized version of this experience.

A laterally pruned ash or elm tree reaches a height of 20–30 metres. To climb and cut the branches from these kinds of trees required both skill and courage. Therefore it is not surprising that this special kind of work was ritualized and linked to a mythological world (cf. Bradley 2005).

The social settings of these rituals are hard to grasp. However, it is worth considering that the people wearing masks with horns were not common people, but operated in a wider social context. A possibility is that these rituals were carried out on the first day of leaf collection each year. This phenomenon is known from agriculture; in certain cultures the making of the first furrows in the ground by an ard was ritualized (Glob 1951).

The rituals depicted on the rock art were therefore probably not carried out on the individual farm by people in care of animals but by certain ritual specialists; this is underlined by the very position of these sites. Both Runohällen and Aspeberget, according to Johan Ling's analysis, are situated in strategic positions in the landscape – they were nodal points where people coming by ship from different directions could meet and assemble (Ling 2008: 127–132, 136–140).

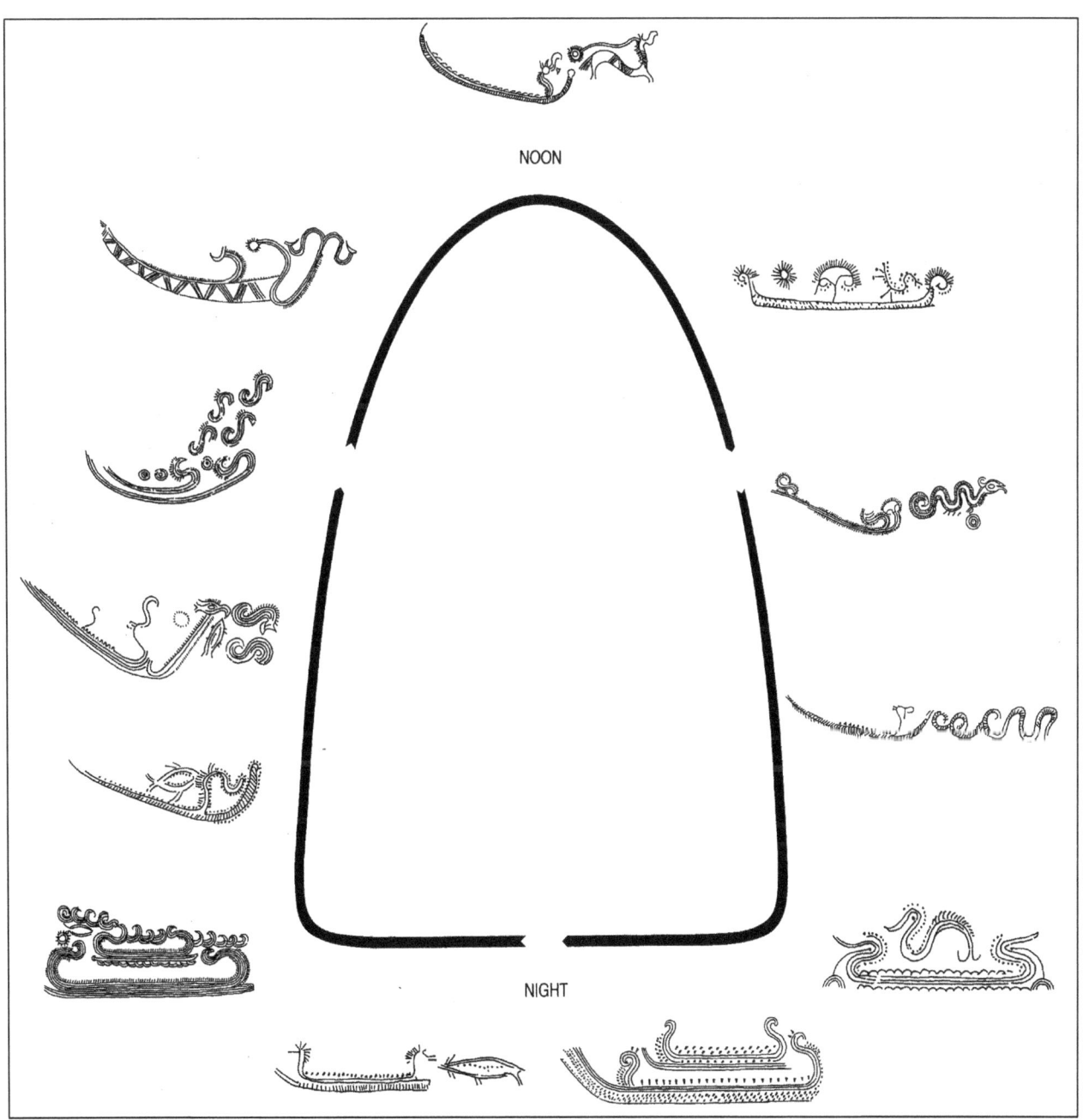

Figure 13. The cyclical journey of the sun inferred from motifs on Danish metal razors from the Late Bronze Age. From Kaul 1998: 262

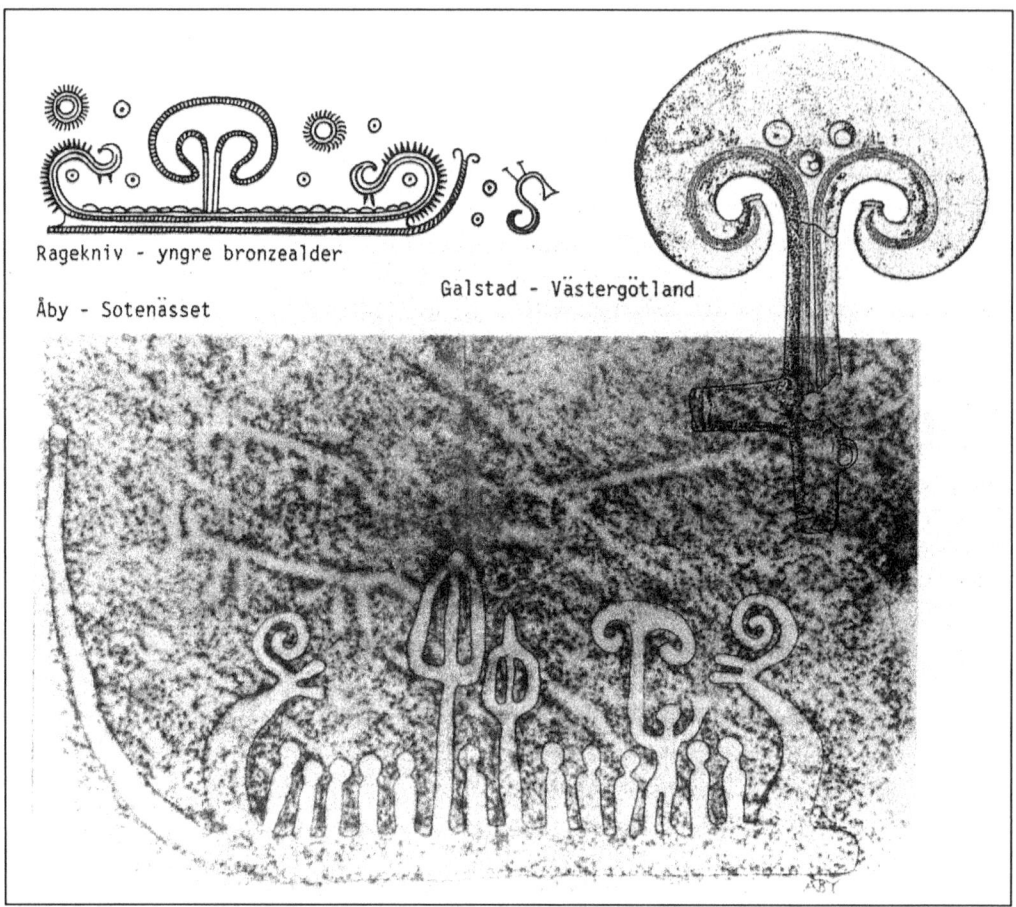

Figure 14. Comparisons between mushroom-shaped symbols on a Danish metal razor, a rock carving from Bohuslän and a cult axe. From Kaul 1998a: 191

TREES AS MEDIATORS BETWEEN EARTH AND HEAVEN

There is yet another body of material that can shed light on the role of the tree in the myths of the Scandinavian Bronze Age, namely the metal items. Especially in the Late Bronze Age there are numerous motifs decorating bronze items such as razors, hanging bowls and neck rings. These motifs somewhat resemble those on rock art but in certain respects they differ.

By analysing the motifs on the razors Flemming Kaul was able to present a new view of Scandinavian Bronze Age religion in a thought-provoking study emphasizing its cyclical character. It turned out that the motifs on the different razors partly overlapped, making it possible to link them together in a narrative focusing on the sun.

On the razors a ship is often seen, which could be combined with sun images and animals like fish, birds, horses and snakes. The sun is transported on different ships and the animals all have their specific task to fulfil in order to help the sun move across the firmament. In the morning the fish and bird are active helping the sun in and out of ships, at midday the sun is drawn by a horse, and in the afternoon the snake is active in helping the sun aboard the night ships which take it below earth to the point of sunrise (Figure 13).

In association with the ships there is, besides from the sun symbols and animals mentioned above, another feature which differs from the others because it does not seem to play any active role in relation to the sun. This element has the shape of a mushroom and in the literature it is sometimes referred to as the mushroom-shaped symbol (Figure 14).

Over the years there have been various ideas about what the mushroom-shaped symbol represents. A minimalistic view put forward by Althin is that it is just an arbitrary sign or ornamentation (1945: 218). Almgren proposed that it could be understood as a tree (1927: 18), a view shared by Gelling and Davidson in their book on the symbols and rituals of Scandinavian Bronze Age (Gelling and Davidson 1969: 59, 132).

Another approach is taken by Kaul, who argues that it should be understood as a representation of a cult axe without a shaft. This conclusion is based on the striking similarities between the mushroom-shaped symbol and non-shafted cult axes (Kaul 1998a: 191). In favour of this idea he draws our attention to rock art where this symbol is in the hand of a human and concludes that this demonstrate that cult axes without a shaft were applied in rituals.

The main problem with identifying the mushroom-shaped symbol with a cult axe is that the axes in that case lack shafts. This is problematic as we have clear evidence from other rock-carvings of shafted cult axes carried around in rituals. The idea of bearing non-shafted axes in rituals seems very much to be a modern view, inspired by the way we see the axes displayed in the museums today. Moreover, the mushroom-shaped symbol is constantly displayed horizontally with the edge turned upwards, so proper axes on the rock art are depicted vertically which is logical if the axe is to be put to practical use (Figure 15).

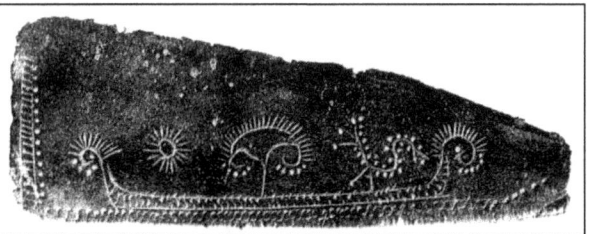

Figure 16. Razor from south Jutland with an elaborate tree in a central position aboard the engraved ship. Kaul 1998b: no. 139

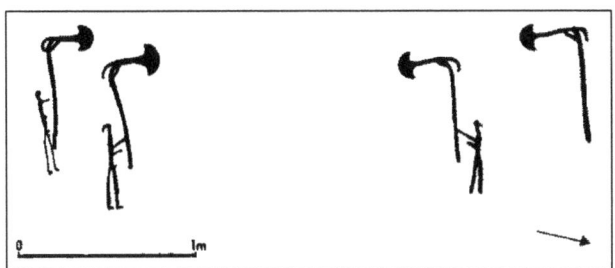

Figure 15. Four men bearing cult-axes at a rock-art site in Simris, Scania. Note the position of the edge in comparison to figure 14. From Skoglund 2005: 112

It should be noted that the mushroom-shaped symbol also occurs on cult axes which might favour the interpretation of this symbol as an axe. However, the occurrence of the symbol on the axes would also be logical if we presume the symbol to be a tree. In that case the symbol would remind people of the practical use of axes when, for example, collecting leaves from trees.

Almgren only stated in general terms that the mushroom-shaped symbol was a tree, but looking at this again, against the background of the discussion of coppiced trees, we can make some interesting observations. In contrast to laterally pruned trees that take on a pole-like shape, pollarded trees take on a shape that resembles a mushroom with a low semicircular or rather flat crown and branches hanging down at its outer ends. Schematic illustrations of pollarded trees sometimes have an outline similar to that of a mushroom (Figure 6).

The mushroom-shaped symbol occurs on 12 Danish razors and all of them except one are highly stylized. The one that differs is a razor from south Jutland depicting an elaborate mushroom-shaped symbol that can be to identified as a tree (Figure 16). Here we see a vertical line dividing into two horizontal lines thus forming a stem splitting into two larger branches. Above and attached to the presumed larger branches is a semicircular line indicating the outline of the crown from which transverse strokes emerge, probably signifying individual twigs. Very close to the bottommost part of the crown on either side of the stem there are five dots, most likely symbolizing nuts or berries (Kaul 1998b. cat. no. 339).

Identifying the mushroom-shaped symbol with a tree enables us to put the tree in a more central position in Bronze Age mythology than has been recognized before.

The tree obviously had some kind of function in the mythology revealed by Flemming Kaul's analysis and interpretations of the Danish razors. As Kaul states, the mushroom-shaped symbol appears on day ships in connection with sun symbols, demonstrating that the symbol should be connected to the day and not the night (Kaul 1998a: 188). Without pushing the material too far, however, it seems as if the mushroom symbol can be related to a narrower spectrum of the day, namely the morning and the evening.

On four out of twelve razors the mushroom-shaped symbol is connected to animals (Kaul 1998b: nos. 132, 275, 339, and 389). In one instance the symbol is associated with a bird and in the other three examples the symbol is seen in relation to horses and snakes or an animal that combines features of horses and snakes. On two razors the angle of the horse indicates that it has just landed on the ship in order to deliver the sun to the evening ship (Kaul 1998b: nos. 339, 389). We may therefore presume that the tree had a role to play in the morning when the bird was active and in the evening when the horse over handed the sun to the snake. The tree thus functioned as a mediator between earth and sky – a consistent idea since the roots of a tree are firmly attached to the earth while the crown strives towards heaven.

TREES AND LIFE-CYCLE RITUALS

Accepting that the mushroom-shaped symbol represents a tree enables yet another intimate connection between man and trees in the Scandinavian Bronze Age. The spectacle fibula (Swedish *glasögonfibula*) from the Late Bronze is usually depicted with non-figurative motifs such as bands, spirals and circles. On a few of these objects it is possible to distinguish motifs on the back of the fibulae – i.e. motifs that were not intended for public display.

In contrast to razors, the motifs on the back of the fibulas are limited to a narrow range of motifs. According to Andreas Oldeberg's survey of the Scandinavian material, the following motifs occur in some numbers on the back of fibulas: hand motifs, the mushroom-shaped symbol and crosses ending in three branches (cf. Oldeberg 1933: 134–135, 190–191, 200). Finally, in Oldeberg's survey there are three motifs resembling a tree similar to those trees depicted in rock art – i.e. pole-like trees with branches coming out of the stem (Oldeberg 1933: 136, 147, 150, 149, 153).

It is interesting to note that the both mushroom-shaped symbol and the pole-like rock-art trees with branches coming out of the stem are represented on the back of the fibulas. If, as has been argued, they both represent trees it is logical to find them in the same context (Figure 17, 18).

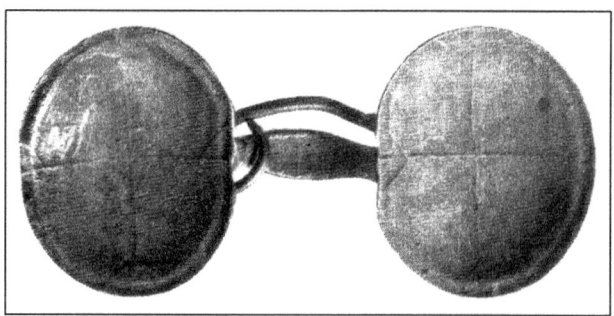

Figure 19. Figure on the back of a spectacle fibula from Kareby, Bohuslän, ending in three branches. From Oldeberg 133: 134

Figure 17. Mushroom-shaped symbols on the back of a spectacle fibula from Repplinge, Öland. From Fredell 2003: 188

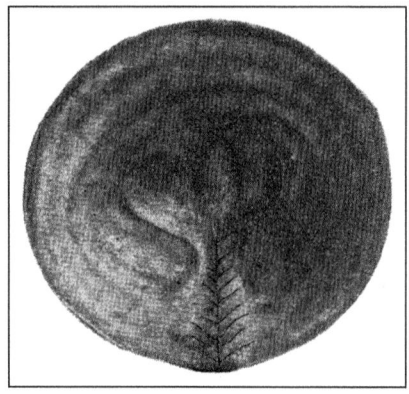

Figure 18. Tree on the back of a spectacle fibula from Repplinge. From Oldeberg 1933: 153, adapted by the author

Perhaps the cross ending in three branches has a connection to trees as well (Figure 19). We recognize this feature from the abstract way of depicting branches and leaf-stacks on rock art. The circular shape determined by the shape of the fibulae and the four lines ending in three branches were elements used to demonstrate branches arranged as leaf-stacks in the rock art of Bohuslän, as discussed earlier. That there might be a similar background for the motifs found inside the fibulas is corroborated by the fact that the cross motif ending in three branches is very clearly concentrated in Bohuslän and the adjoining provinces in Sweden and Norway (Oldeberg 1933: 145).

The mushroom-shaped symbol on the inside of the fibulae sometimes clarifies a pattern that is not obvious when one looks at the outside of the fibulae. Here one can see bands ending in two spirals that have an arbitrary meaning to an observer. However, on the inside the mushroom-shaped symbol is the positive imprint of the backward bent spiral – i.e. the mushroom-shaped symbol – giving it a more specific meaning for the owner than for the public (Oldeberg 1933: 139, 177; Fredell 2003: 188).

The motifs on the back of the fibulas are generally not engraved but were made by the bronze smith simultaneously with the object. The very limited but varied numbers of motifs hidden for public indicate that they were designed with specific people in mind. Following Oldeberg, these fibulas did not only have a functional purpose but they were also symbols with a protective code functioning like amulets (Oldeberg 1933; Marstrander 1963; Fredell 2003: 188).

The fibulas were applied to female dresses. Unfortunately, we do not know very much about how these items were applied to the dresses since in the Late Bronze Age people were cremated with only a few objects put into the grave. The majority of fibulas are stray finds or from deposits, but before being deposited they were personal items applied to the dress.

We might get a glimpse of how these fibulas were worn when looking at the drawing of the Grevensvænge find. This find from Zealand in Denmark was found in the 18th century and consisted of six bronze figurines that originally were probably applied to a wooden object. Unfortunately, the find was split and several of the figurines were lost, but on a drawing from 1779 we can see a woman with a fibula placed in a horizontal position at, or just above, the breast (Djupedal and Broholm 1953; Glob 1962).

Judging from this drawing, the fibula had a prominent position on the female dress, and in accordance with the interpretation of some other bronze objects like the razor we may assume that it was given to the woman as part of a life-cycle ritual (Arcini et al. 2007: 143–144).

TREES AND CREATION MYTHS

In Old Norse mythology it is said that two tree trunks were discovered upon a sea shore from which the first humans couple were formed – Ask (male) and Embla (female), usually taken to indicate ash and elm respectively. In the legends the ash was connected to power resident in water, a situation that is also found in Greek mythology where the ash tree was sacred to Poseidon, who was also the god of horses (Winterbourne 2004: 77).

In Scandinavian mythology the ash tree *Yggdrasil* had a central position. The tree had three roots stretching out over the world, and below them was the world of the humans, the giants and the kingdom of the dead – Hel. The tree thereby connected heaven, earth and the underworld.

Yggdrasil is also related to the Norns – female beings connected to the idea of fate. One of their duties was to take *aur,* a kind of mud, from the well of Urd and to pour it on the roots of Yggdrasil so that it would not dry (Andrén 2004: 390–394). There was thus a connection between the ash and water in the mythology of late Scandinavian prehistory.

An association between trees, water and horses also seem to be found in the Bronze Age material. Looking at the rock art from this perspective, it is obvious that there was an ambiguous relationship between human, ships, horses, snakes and trees. Many of the ships depicted in rock art, and on razors, end in a horse's head and sometimes we found creatures that combine features of horses and snakes. Aboard there are sometimes one or two trees standing and we also have a case where the stem of the ship ends with a mushroom-shaped symbol – in this study interpreted as a tree.

The panel from Aspeberget discussed above is intriguing in this context. Here we interpreted the three-folded object as a branch in the hands of a human, feeding a horse's head attached to the stem of ship. This makes it a mythological ship partly made of living animals and in need of constant feeding with branches if it is not to die. Considering the very close spatial relationship between ships and trees – with trees standing aboard ships or being depicted just beside ships, or sometimes overlapping the ships, one can ask whether the ship was not also considered partly as a tree.

These perspectives give us yet another idea of how to understand the combination of trees and horses aboard ships. When trees are aboard ships we may understand them as growing out of the ship – this is especially clear when the mushroom-shaped symbol is attached to the stem – in these cases there is no clear division between the ship and the tree but they are unified. If – as has been argued before – the ships were ritualized and regarded as partly horses and partly ships and thereby in need of fodder; the ship-horse could actually feed itself by eating from the branches hanging down from the tree aboard the ship. The combination of trees and horses occurs on several of the Danish razors with tree depictions (Kaul 1998b: nos. 104, 132, 275, 330, 335, 339, 371, 389, 390) and it also occurs on rock art.

A razor probably from Lolland in Demark is revealing (Figure 20). Here we see a ship folded by the engraver to maximize the use of space. At the prow there is a stylized animal's head, probably a horse, with its head close to – and directed towards – a mushroom-shaped symbol growing out of the keel extension. The head is directed towards the tree as if it is about to eat from the crown of the tree (Kaul 1998a: 225–226).

The Hjortspring ship, dated to about 400 BC, or the very earliest part of the Scandinavian Iron Age, is the oldest preserved Scandinavian ship. It was made out of five planks from lime trees up to 15 metres long and 0.7 metres wide. The planks were thereby made of a tree measuring 1 metre in diameter and with a straight stem of 16 metres length (Crumlin Pedersen and Trakadas 2003). The lime tree is known to be well suited for pollarding and pruning in order maximize the production of branches; and the lime branches are valued as animal fodder.

The preserved ships from the Scandinavian Iron Age are mainly made of oak, which has many good qualities as timber for building boats, but oak branches are less valuable as a source of fodder than lime. The preference for oak is partly due to technical developments during the course of the Iron Age. Improved axes facilitated the making of planks by the radial splitting technique, which required larger trees like oaks (Larsson 2007: 85–91). Assuming that the choice of wooden planks in the older

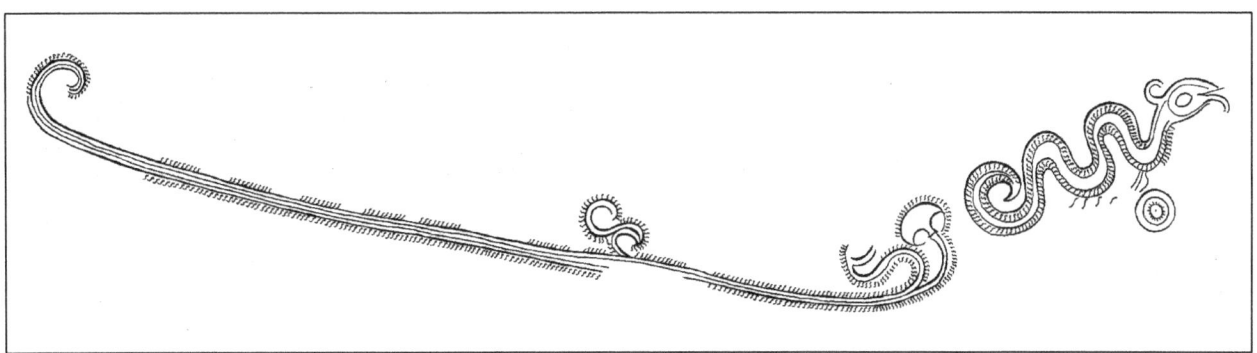

Figure 20. Razor probably from Lolland, Denmark. At the prow there is a stylized animal's head, probably a horse, close to a mushroom-shaped symbol. From Kaul 1998b: 226

Figure 21. Hanging bowl from Öland, south-east Sweden. From Kaul 1998a:192

Hjortspring find is not unique, we can imagine that in the Bronze Age ships were made out of trees that earlier in their life history had been coppiced and used as sources for fodder.

Trees growing out of ships can also be studied on some of the hanging bowls. Trees – in the shape of mushroom symbols – are quite often found on hanging bowls from the Late Bronze Age, where several of them occur as a circular running band. On rare occasions the quite abstract decoration on the hanging bowls turns more solid, and we can distinguish features similar to those on the razors.

This is the case on a hanging bowl from Öland, south-east Sweden. In the centre of the bowl there is a circular symbol that in this context ought to represent the sun (Figure 21). The sun is surrounded by spirals that end either in horse's heads or snake's heads. As Kaul has pointed out, a majority of the figures have lines in front of their bodies that very much resembles a keel extension often seen on the ships on razors and in rock art. The ships are therefore most probably represented as well. On two of the horse-snakes' bodies there is a mushroom-shaped symbol attached (Kaul 1998a: 192).

On the hanging bowl from Öland we again find the combination of trees, ships horses and water; moreover, the trees occur as a pair of trees. It is worth considering that the mushroom-shaped symbol, like other kinds of depictions of trees, either occurs individually, or paired, but never in higher numbers either on razors or in rock art. This hints that there might have been two different kinds of trees that represented two different principles. In later Scandinavian mythology we know from written sources that the first man and woman were believed to have been created out of an ash and an elm respectively. Judging from the hanging bowl from Öland, it seems reasonable to conclude that a similar idea might have existed in the Bronze Age as well.

The association between humans and trees is a worldwide metaphor (Rival 1998b). This could be expressed in many ways, but one way of elaborating upon this is the idea of the body as a stem and the legs and arms as branches. Perhaps the most convincing evidence of the existence of this idea, also in the north-European Bronze Age, is a motif on a bronze sheet from northern Germany. On this sheet, dated to the Late Bronze Age, we see a human aboard a ship with the arms raised – branches are growing out of the body, transforming the human into a tree or alternatively a human being growing out of a tree (Figure 22) (Sprockhoff 1956; Fredell 2003: 241).

The human tree is put in a cosmological setting aboard a ship that is surrounded by sun symbols from which snakes with horse's heads originate – or are being born. At the bottom of the sheet we see two trees opposing each

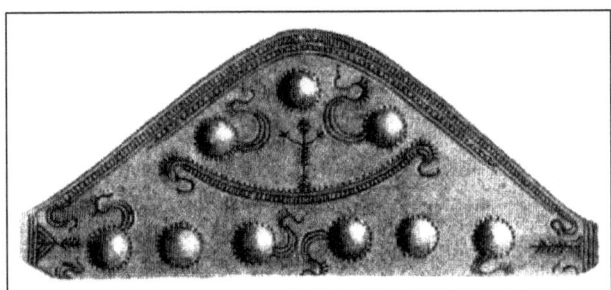

Figure 22. Bronze sheet from northern Germany. From Fredell 2003: 241

other. The duplicated trees at the bottom of the sheet and the transformation of a tree into a human – or vice versa – seem to demonstrate that there existed an idea of humans being born out of trees already in the Bronze Age.

A LONG-LASTING PRACTICE

Because of the central position of the tree in Old Norse mythology, the general lack of evidence for tree symbolism in the archaeological record has been somehow difficult to understand. This situation has recently been changed by the work of Anders Andrén, who argues that the Iron Age stone settings in shape of tricorns dating to 200–1050 AD should be understood as representations of trees where the three arms coming out from the centre of the monument are homologous to the roots of the *Yggdrasil* tree (Andrén 2004: 407–418). Moreover, the shape of the tricorn is duplicated on shield and shield buckets dating to 550–800 AD; and the shape was reproduced on the outline of women's brooches dating to 800–1050 AD (Andrén 2004: 418–421).

From the earlier parts of the Iron Age, 500 BC to 200 AD, evidence of tree symbols in Scandinavia is not so common. Nevertheless, there are triangular monuments that predate the tricorn and might symbolize a more stylized version of a mythological tree. These kinds of monuments generally date to the period 500 BC to 550 AD (Andrén 2004: 411).

There is one observation indicating that there was some kind of consistency in the way the tree was conceptualized from the Bronze Age to the Iron Age. As Andrén clearly demonstrates, tricorns occur individually without connection to other graves, or they are just represented by one or very few examples in Iron Age grave-fields. On the grave-field they often take a prominent position as if they are among the oldest monuments on the site. Moreover, about one third of them lack burials (Andrén 2004: 408). This makes them different from the surrounding monuments like mounds and ship settings, which often occur in large numbers. In contrast to the ships, the ideas connected to trees did not need to be duplicated by monument being added to monument; their mere presence as a single monument was sufficient at the grave fields. The situation on the Bronze Age rock-art panels is strikingly similar. While ships are duplicated in impressive numbers, the tree occurs alone or in pairs.

Scandinavian mythology had many origins. Some of these were recent while other elements had been in use over a substantial length of time. The ritual practice and the ideas connected to trees were constantly changed to fit into people's new experiences. On a general level we may recognize ideas like the tree as a mediator between earth and heaven, and the idea of a very intimate relationship between people and trees both in the Bronze Age and in the Iron Age; but when we go into details the differences begin to appear (Andrén et al. 2006; Andrén 2006).

Though we can distinguish some similarities between the tree in the Bronze Age iconography and in the Iron Age mythology, this does not mean that similar ideas existed over time. However, the constancy with which the relationship between single trees and many surrounding ships was played out in Bronze Age rock art and at the Iron Age grave-fields underscores that certain ideas might have prevailed for centuries or even millennia.

Chapter 3.
THE ORDERING OF TREES

A lesson learned from the previous chapter is that Bronze Age trees were reshaped by man to fulfil specific purposes – i.e. they were culturally modified. Rock-art images have a great potential in bringing us knowledge about man's interaction with specific trees, but they do not inform us about the shape and character of the woodland. How was the overall landscape planned and maintained in order to produce trees with certain values?

One way to approach these questions is to look for analogies among societies known from written sources. The debate in North America is particularly revealing. From literature on the use of fire among North American native people it is clear that landscape management was linked to the use of fire. Fire was used to remove weeds after fallows, to clear land for the arrangement of new agricultural plots and to facilitate hunting. The use of fire shifted from area to area. In some areas fire was used on a regularly and yearly basis, while in other areas fire was rarely used (Day 1953: 334–339; Pyne 1982: 71–83; Cronon 1983: 48–52; Whitney 1994: 107–120; Foster *et al*. 2004: 62–72).

Regular use of fire transformed the forest to woodlands with densely spaced trees. The use of fire also created conditions favourable to strawberries, blackberries, raspberries and other gatherable fruits (Cronon 1983: 51). The trees close to settlements were used for fuel and berry collection, while woodland at greater distances was used for small game and foraging (Williams 1989: 40). The results of some studies indicate that the overall distribution of the trees in the landscape was affected; different zones in the landscape were characterized by slightly different tree compositions (Foster *et al.* 2004).

There are many different opinions about the extent of the fire regimes used by Native Americans and to what degree the forest was affected by fire. However, there seems to be a general consensus that the various kinds of agricultural systems and woodland practices present at the time of European settlement cannot be understood without considering the role of human-induced fire (Barret *et al*. 2005).

This situation is very different from the discussion of Bronze Age and Iron Age agriculture in northern Europe. In this context agriculture is generally approached from two lines of thought – the management of cattle and the production of crops. Issues such as the introduction of keeping cattle in byres, the composition of the stock or how certain changes in agricultural tools improved the rate of return are regularly discussed, but little attention is paid to the use of fire and trees.

The discussion of fire and the role of trees in connection with agriculture going on in North America thus addresses a request to European archaeologists to incorporate trees and fires into their models of prehistoric agriculture. Of course this cannot be done by simply taking over ideas since there are obvious differences in the way agriculture was practised between the two continents.

The most obvious difference is that the agriculture of the native people of North America did not involve domesticated animals. The woodland was therefore not affected by domesticated and grazing animals, which made fire even more important in the clearing of vegetation compared to Europe. The occurrence of domesticated animals in Europe implies yet another important difference between the two continents – in Europe trees was not only important as producers of nuts, fuel and timber but they also could supply fodder through pollarding and coppicing.

Irrespective of these differences, the discussion in America seems relevant in bringing the role of fire and the structured utilization of trees into the core of the agricultural debate. From this perspective there seems to be an obvious need to broaden the discussion of trees and fire as important ingredients in prehistoric landscape analyses. In the following this will be done by considering the record of charcoal from the Småland Uplands.

CHARCOAL – RECORDS FROM BOGS

Recent palaeoecological research in the Central Småland Uplands has provided results that have a bearing on the discussion of the use of fire in prehistory (Greisman and Gaillard 2009; Olsson and Lemdahl, 2009; Olsson *et al.* 2010; Marlon *et al.* 2010). The results from the Stavsåkra bog are of special interest. The site is situated 500 metres from the lake Helgasjön and is centrally located in one of the two study areas that will be discussed below in this chapter.

Marie-José Gaillard-Lemdahl, Geoffrey Lemdahl and their team (at Linnaeus University, Kalmar, Sweden) applied a broad spectrum of palaeoecological methods to the peat profile from the Stavsåkra bog that covers the entire Holocene period from c. 9000 BC until today. A combination of analyses, i.e. charcoal (microscopical < 160 µ and macroscopical > 250 µ), pollen, plant macrofossil (e.g. seeds, fruits, leaves) and insects remains (coleoptera in particular), led to new and partly controversial results concerning the use of fire in the shaping of the Bronze Age and Iron Age cultural landscape.

The amount of microscopic charcoal increases significantly from c. 1000 BC and the values remain relatively high until c. AD 1900. The occurrence of macroscopic charcoal and coleoptera species depending upon regular fires for their existence during the period 1000 BC–AD 1800 confirms that regular fires took place in the area around Stavsåkra. Since there is no known climatic change around 1000 BC in southern Scandinavia that can explain the change in fire intensity, it was concluded that these fires were human-induced (Olsson *et al.* 2010).

Moreover, high abundance of pollen from *Calluna vulgaris* (heather) and the occurrence of coleoptera species dependent on large areas of *Calluna* heaths testify to the formation of regularly burned *Calluna* heathland in the vicinity of the Stavsåkra bog from c. 750–500 BC. We know from historical documents and many palaeoecological studies in Europe that *Calluna* heathlands were used by humans as grazing land, and that they were maintained by the regular use of fire to rejuvenate the *Calluna* bushes for better fodder (Greisman, 2009). It is suggested that the timing of the formation of *Calluna* heaths at the study site is related to soil impoverishment due to the intensive use of fire from c. 1000 BC, and to several centuries of grazing from the Late Neolithic/Early Bronze Age, as shown by the results of the pollen analysis. Therefore, it was proposed that the area was characterized by grazing land dominated by *Calluna* and managed using fire from 750 BC until the 18th/19th century (Greisman *et al.* 2009, unpublished, in Greisman 2009).

In conclusion, the use of fire was more common and occurred on a larger and more regular scale in the cultural landscape of the Bronze Age and the Iron Age than in the landscapes of the Neolithic. It is difficult to estimate the geographical relevance of the results from the Stavsåkra bog study. However, from an archaeological perspective, Stavsåkra is by no means unique, but is rather a standard site in this part of Sweden.

Therefore, in the forthcoming discussion, the results presented above are proposed to be relevant for the 20 x 30 kilometre large area around the lake Helgasjön which is defined as one of the study areas to be discussed in this chapter.

CHARCOAL – RECORDS FROM ARCHAEOLOGICAL CONTEXTS

Sampling of charcoal from archaeological contexts for ^{14}C dating is a standard procedure. If the charcoal is identified to species level before the analysis, the value of information will increase. For example, if a piece of charcoal is identified as oak and dated to the Early Roman Iron Age this does not only inform us that wood grow in the area in those days but also that the wood by some reason burnt. The context of the charcoal will give us hints of the character of the fire. If the charcoal was taken from a hearth it might have been a fire for cooking and lighting, if it was taken from a clearance cairn it might have been a fire for clearing vegetation, and finally if it was taken from a post-hole in a long-house it might have come from a house that burnt down (Lagerås 2000a: 198–199).

The possibilities of using species-identified charcoal from archaeological contexts was demonstrated in a Scandinavian setting by Per Lagerås in a study based on materials from the Hamneda project in the Western Småland Uplands. In his study wood identified charcoal was used to ascertain different phases of wood clearing in the Late Bronze Age and Early Iron Age. The initial clearing was characterized by the removal of oaks, while the later clearing was characterized by the clearing of birch and hazel (Lagerås 2000a; Lagerås and Bartholin 2003).

The upcoming discussion draws heavily on Lagerås's pioneering study, but it includes a larger sample compiled from two different areas in the Småland Uplands: the Western Småland Uplands (the Hamneda area) and the Central Småland Uplands (the Växjö area). The data concerning ^{14}C dates from these and other excavations in the region (Kronoberg County) were stored in a database at Småland Museum. The database contains information on the site, the find context, ^{14}C date and when applicable the species identification of the charcoal.

From the overall database a sample of ^{14}C-dated charcoal pieces from the Western Småland Uplands and the Central Småland Uplands was compiled (for definitions see below). ^{14}C-dated charcoal pieces that were not identified to species were sorted out, and the same goes for dates that were younger than the Viking Age or older than the Bronze Age.

Moreover, if the dated charcoal pieces could not be connected to a prehistoric context they were taken away from the sample. This happened in a few cases when features at medieval sites had been given prehistoric dates. There were also a few examples of species identifications with a question mark and of some samples assigned to different species; these too were withdrawn from the sample.

When these procedures was carried out there was a total of 252 species-identified charcoal pieces dating from the Late Early Bronze Age to the Late Iron Age that could be attributed to a specific prehistoric context; 167 of these were from the Central Småland Uplands and 85 of these were from the Western Småland Uplands.

The 167 and 85 samples from the two areas were arranged in chronological order using the BP values and divided among the following contexts: clearance cairns, settlements and graves. It should be noted that the category of settlement is a broad group including hearths, cooking pits, pits, post-holes, heaps of fire-cracked stones and ovens. They reflect domestic activities in a very broad sense, and it is not stated whether they were part of a settlement with permanently used houses, or if they represent occasionally used camp sites.

Finally, certain characteristics – such as the changes in the relative proportions of tree species or the occurrence of certain tree species in new archaeological contexts – were used to further subdivide the tables to make them reflect specific periods in landscape history.

There are obstacles in using this material in order to draw conclusions about the prehistoric use of wood. A difficulty when using charcoal from archaeological contexts in interpretations is that different species behave differently when burnt. The burning of birch will result in brittle charcoal that easily gets fragmented, while oak is more solid. The collection of visible charcoal pieces might therefore lead to the overrepresentation of certain species such as oak (Lagerås 2000a: 200).

However, when selecting charcoal for radiocarbon dates there is a wish to minimize the affect the age of the wood has on the ^{14}C results. There is an endeavour to choose charcoal that represents young wood since the radiocarbon dating will date the age of the wood and not the age of the burning incident. Therefore oak might be neglected if charcoal from, for example, birch – which is generally less old than oak – is present.

This is obvious when studying the results from the excavation at *Rottnevägen* north of Växjö in the Central Småland Uplands, which is listed in table 1. The clearance cairns at this site are dated to the period 1600 BC to AD 1700. The species represented in the table therefore give an accumulated picture of the species burning at the site during roughly 3,300 years of clearance.

Birch that is well represented in the sample is also well represented in the dated material, but even though oak occur in the record from Rottnevägen none of these

Table 1. Species-identified charcoal from clearance cairns from Rottnevägen. One sample with a mixture of aspen, birch and oak is not listed in the table. Data from *Smålands museum rapport 2003*: 60

	Occurrences	No.	No. dated
Birch	31	69	14
Oak	6	19	
Aspen/*Salix*	2	2	1
Asp	2	2	
Salix	1	1	
Deciduous tree	2	3	1
Spruce	4	14	1
Juniper	1	2	
Pine	13	23	2
Conifer	1	1	1
	63	136	20

pieces were ^{14}C-dated. The latter is due to the selection process when sending samples to the laboratory.

Another thing to consider is the sampling procedure which is usually quite similar within each context but may vary between contexts. Charcoal beneath clearance cairns is conventionally sampled individually, while charcoal from settlements sometimes is taken in larger samples.

The conclusion is that we cannot presume that the composition of species-identified and radiocarbon-dated charcoal from a certain site reflects the composition of the trees in the surrounding cultural landscape. This is because decisions made by archaeologists will affect the results.

However, there is little reason to believe that the sampling practice or decisions of the archaeologist change depending upon the period or region in focus. If a pattern from a certain period or region is compared to other periods and regions and significant changes in the material can be identified, the question should be posed whether these patterns are the outcome of people's practice in prehistory. Having these considerations in mind it should be possible to discuss man's interaction with trees using species-identified charcoal. In the following this will be done by considering two different areas in the Småland Uplands in southern Sweden (Figure 23).

THE USE OF TREES IN THE CENTRAL SMÅLAND UPLANDS

The Central Småland Uplands is in this study defined as an approximately 20 x 30 kilometre large area around the lake *Helgasjön* and the city of Växjö, Sweden. The lake Helgasjön covers about 50 km^2 and is centrally positioned inside the studied area. The landscape is characterized by

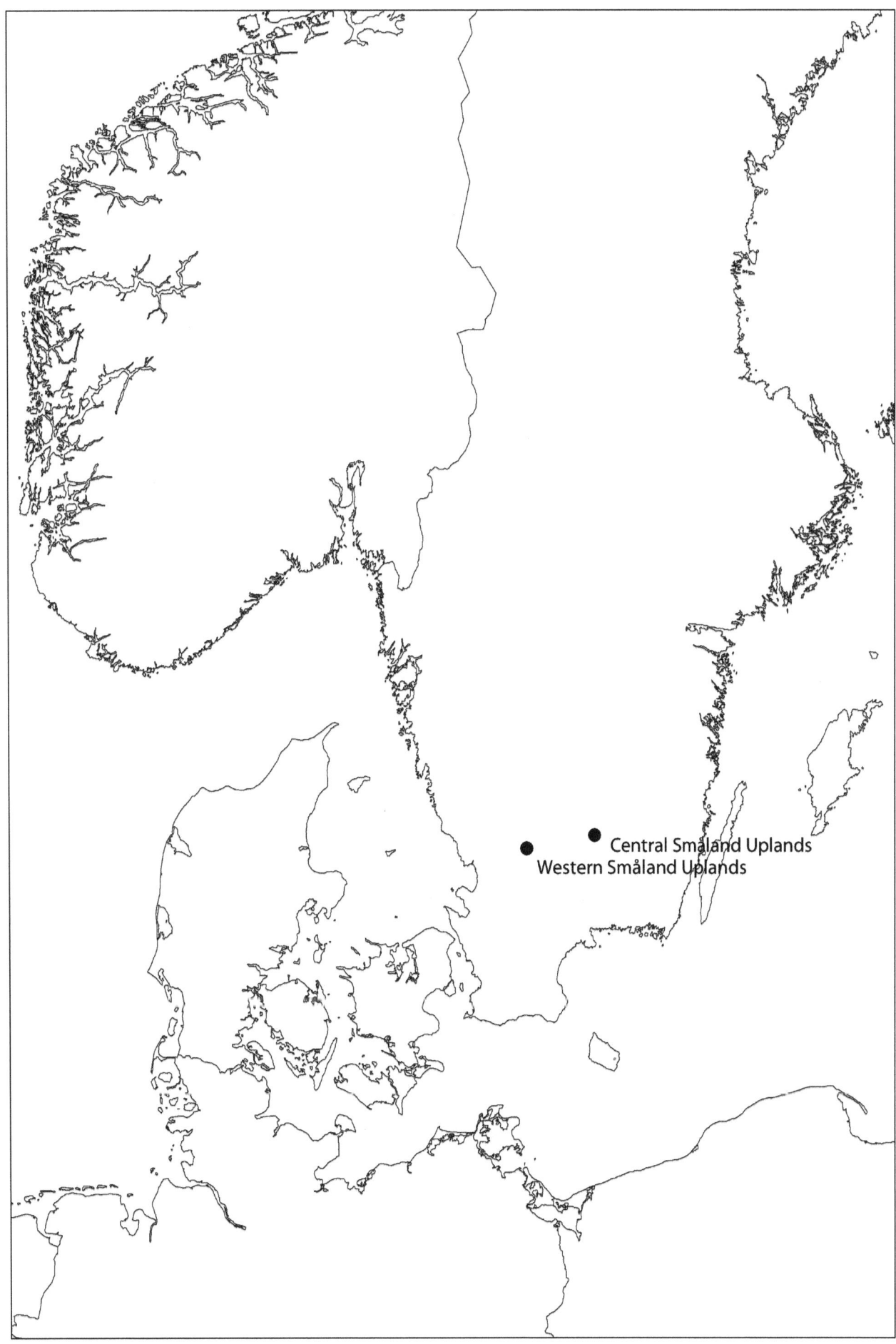

Figure 23. Map showing the location of the Western and Central Småland Uplands in Sweden

drumlins oriented north–south, which have structured landscape use at least since the Bronze Age.

The identified and dated charcoal pieces from the Bronze Age, 1700–500 BC, are presented in table 2. This period is characterized by the introduction of clearance cairns which took place at the very beginning of the period. The 62 pieces of charcoal dated to this period are distributed among three contexts: clearance cairns (19), settlements (30) and graves (13) (Figure 24). However, the different species are not evenly spread among these three categories; some species are over-represented in certain contexts.

Table 2. Species-identified charcoal from the Bronze Age (3430–2340 BP), distributed by context

	Clearance cairn	Settlement	Grave	No.
Alder	1	9		10
Birch	13	16	5	34
Oak	2	2	2	6
Lime	1		1	2
Maple	1		1	2
Pine	1			1
Elm		1		1
Ash		1		1
Salix		1	1	2
Pomidae			1	1
Rowan			1	1
Hazel			1	1
	19	30	13	62

Figure 24. Clearance cairn, Tjureda Parish, Central Småland Uplands. Photo: Peter Skoglund

Birch totally dominates among the dated charcoal pieces from clearance cairns while oak, lime, maple and alder occur in low numbers. Birch also dominates the picture at the settlements, but alder is also represented in high numbers, while oak, elm and ash occur in low numbers.

In the grave context there is a very wide range of trees represented in small numbers. Here we find birch, oak, *Salix* and maple, which are also represented in other contexts; but we also find lime, pomidae, rowan and pine that are restricted to grave contexts only.

Different kinds of trees were treated differently. Alder and birch were in a sense bulk trees that were taken in larger numbers from the surrounding landscape to the settlements where they were used for fuel. The 30 charcoal pieces of alder and birch from settlement contexts were found in hearths (18), cooking pits (6), pits (2), post-holes (3) and a heap of fire-cracked stones.

Even though alder and birch were used in a similar way at the settlements, they came from two different areas in the landscape. Alder prefers a wet environment and establish itself primarily at, or close to, rivers, lakes, bogs and swampy areas, while birch has a more general distribution in the landscape.

Thirteen of 19 charcoal pieces from clearance cairns are from birch, which makes birch the most frequently represented tree in this context. Because of the preference for young wood when sending the samples to the ^{14}C laboratory, oak is probably underrepresented in the material, which is indicated by a comparison between tables 1 and 2, but also by results from pollen analysis.

From the pollen analysis carried out at Stavsåkra we know that a major clearing of primarily oak and lime took place during the Early Bronze Age, c. 1500–1100 BC, resulting in higher values for birch and hazel (Greisman *et al.* 2009: 16). Because of decisions made by archaeologists, the clearance of oak and lime is only vaguely reflected in the identified charcoal from the clearance cairns: one of the two charcoal pieces identified as oak and the lime charcoal piece fall within the period of major clearing 1500–1100 BC.

However, hazel that was favoured by these clearances is only found in one example in table 2, attributed to a grave context. In contrast to the absence of oak, the absence of hazel below clearance cairns is probably significant. Hazel has a low age, like birch, and there is no reason to think that this specie would not have been used for dating if available. It should be noted that hazel is not available in table 1 which include all wood identified charcoal sampled below clearance cairns at the excavation at Rottnevägen.

A possible explanation to the very limited occurrence of hazel in this record is that relatively few hazels grew in the landscape and that these were gardened and taken care off – i.e. they were not affected by the fire regimes practiced by Bronze Age man. Such a situation might occur if there was a high pressure on the landscape caused by for example grazing cattle something that would not favour the growth of hazel. This interpretation seems reasonable in the light of the distribution of the species in the Iron Age period which is presented in table 3.

Table 3. Species-identified charcoal from the Early Iron Age (2340–1765 BP) distributed by context

	Clearance cairn	Settlement	Grave	No.
Aspen	1			1
Alder	5	14		19
Birch	8	18	5	31
Oak	1	2	1	4
Hazel		3		3
Ash		1	1	2
	15	38	7	60

Table 5. Species-identified charcoal from the Late Iron Age (1760–1005 BP) distributed by context

	Clearance cairn	Settlement	Grave	No.
Birch	8	13	4	25
Oak	1	3	1	5
Aspen	1		2	3
Alder		7		7
Hazel		2	1	3
Ash			1	1
Salix			1	1
	10	25	10	45

From around 500 BC charcoal pieces of birch found below the clearance cairns are supplemented with charcoal pieces of alder. Since alder primarily establishes itself in wet environments, this indicates that clearances for agriculture now took place in low altitudes in the landscape; in close connections to streams, bogs and swampy areas.

Since alder is also represented at settlements one cannot rule out the possibility that these pieces of alder originally burnt on settlements that later on were turned into cultivated land. However, the dating of these alder pieces seems to contradict such a conclusion. The charcoal pieces identified as alder cluster in a limited period of time, which is very different from, for example, birch which occurs below clearance cairns in all three periods. This also goes for alder found at settlements, which occurs in all three periods (tables 2, 3 and 5).

Apart from the five charcoals of alder from this period we only know of a single piece from the Bronze Age dated to 1780–1430 BC (2 sigma). Moreover, four of the five charcoal pieces of alder from the period in question stem from the earlier part of the period: c. 700–100 BC using two sigmas and c. 700–200 BC using one sigma (table 4). The limited age spectrum indicates that we are dealing with a certain kind of landscape management restricted to a limited period of time.

From the pollen analysis at Stavsåkra it has been suggested that alder swamps were cleared for grazing from around 1000 BC (Greisman et al. 2009: 13). During the second century BC there are indications of lower grazing pressure in the pollen analysis. Lime and hazel increased, as did alder, which might indicate that grazing areas in alder swamps were not cleared any more (Greisman et al. 2009: 14). Therefore it is interesting again to note that four out of a total of six clearance cairns dated by charcoal from alder fall within the period 700– 100 BC (2 sigma) or 700–200 BC (1 sigma).

We do not know why the alder at these fens was burnt – one reason could be the collection of alder for fuel – a tradition established in the Bronze Age and continuing into the Iron Age (tables 2, 3 and 5). After alder had been collected for fuel, leftovers on the ground were burnt; some of this did not fall on the fen itself but on dry ground close to the fen and was later on covered by clearance cairns. There is also the possibility that alder fens were cleared for grazing or fodder production (Lagerås 2002: 395–398). Irrespective of why alder was burnt, the occurrence of alder below clearance cairns indicate an intensified use of the cultural landscape.

Compared to the Bronze Age, maple, lime and elm have disappeared from the record. These trees certainly existed in the landscape but their disappearance from this record indicates that they were rarely used in domestic or ritual settings that involved fire. In the more open landscape that was formed during the period 750–500 BC these trees became rarer and those that prevailed were probably considered for specific functions such as fodder production, which would leave no traces in this kind of record.

The record from the Late Iron Age, AD 400–1000, is presented in table 5. The main difference in comparison to the earlier period is the disappearance of charcoal

Table 4. The dating of clearance cairns with ^{14}C-dated charcoal identified as alder from the Iron Age

Place	No.	BP	±	Cal. 2 sigma	Cal. 1 sigma
Räppe	Ua-25786	2340	70	752–206 BC	705–235 BC
Räppe	Ua-25785	2290	75	735–121 BC	409–206 BC
Sånnestorp	Ua-17145	2275	85	735–93 BC	405–203 BC
Hovshaga	Ua-18239	2195	60	392–102 BC	362–197 BC
Hovshaga	Ua-18238	1765	65	90–413 AD	140–397 AD

identified as alder found below clearance cairns. This indicates that areas close to alder swamps were no longer used for agriculture or that the alder fens did not burn anymore. However, the alder swamps were still used to extract fuel, as is evident from the occurrence of alder in hearths at settlements.

We should now turn our attention to the Western Småland Uplands, which will reveal quite a contrasting picture of man's interaction with trees during the Bronze Age and the Iron Age.

THE USE OF TREES IN THE WESTERN SMÅLAND UPLANDS

The Western Småland Uplands are defined in this study as an approximately 15 x 3 kilometre large area west of the river Lagan and south of Ljungby. The landscape is hilly terrain characterized by moraine. However, only a couple of hundred metres to the east there is the Lagan valley consisting of finer sediments. The landscape is very much structured by the river Lagan, one of largest rivers in south Sweden.

The record from the Bronze Age and the Early Iron Age, 1700 BC–AD 100, is presented in table 6. Only two of the charcoal pieces in the table can be attributed to the Early Bronze Age; these are the piece of hazel found in a gallery grave and the piece of oak found below a clearance cairn at the same site.

Table 6. Species-identified charcoal from the Middle and Late Bronze Age and the Early Iron Age (3140–1925 BP) distributed by context

	Clearance cairn	Settlement	Grave	No.
Oak	10	2		12
Lime	1	1		2
Aspen	1			1
Maple	1			1
Salix	1			1
Alder		1		1
Hazel		1	1	2
	14	5	1	20

From around 1000 BC the picture changes and there is continuous record of dates from clearance cairns. A majority of the charcoal pieces found below the cairns are from oak, while lime, aspen, maple and *Salix* occur in low numbers. The dominance of oak in this period demonstrates that clearances by fire took initially took place in a wood with high trees (Lagerås 2000: 201). The dated settlement remains consist of two hearths, a cooking pit and two pits. No houses have been documented from this period in the study area.

The record from the Roman Iron Age and the Migration Period, AD 100–550, is presented in table 7. This period is characterized by the occurrence of proper settlements: at one of the sites a farm consisting of two long-houses replacing each other on the same spot was documented, and at another site outbuildings and associated activity areas were documented (Cronberg et al. 2000).

Table 7. Species-identified charcoal from the Roman Iron Age (1920–1505 BP) distributed by context

	Clearance cairn	Settlement	Grave	No.
Hazel	10	14	3	27
Oak	3	4		7
Birch	13			13
Lime	1			1
Ash	1			1
Alder		2		2
	28	20	3	51

This is also the period to which a majority of the ^{14}C-dated clearance cairns are dated. The use of the trees differs from the preceding period. Oak is only represented in small numbers and instead the picture is dominated by hazel and birch. This demonstrates that the clearances now took place in a different landscape compared to the previous period. The landscape was more open, favouring the occurrence of light-demanding species such as hazel and birch (Lagerås 2000a: 201).

Making a distinction between hazel woods and hazel nuts on different contexts demonstrates that burnt hazel nuts are generally found at the settlements while burnt hazel wood is generally found below the clearance cairns (table 8). Because of the very low age of the hazel nut – 1 year – hazel nuts would have been used for ^{14}C dates if available. Therefore the lack of charred wood of hazel and birch at the settlements is explained because hazel nuts were found at the settlements and favoured for ^{14}C dates because of their low age. In contrast, the absence of burnt hazel nuts below the clearance cairns is significant; if burnt hazel nuts had occurred in these contexts they would probably have been noted and dated.

The appearance of burnt hazel nuts at the settlements is probably due to roasting of nuts, which can be further studied from the archaeological record. At one of the sites an outbuilding measuring 4.0 x 3.3 metres was documented, oriented east–west. The building had no roof-bearing posts but the roof reposed the walls. There was a presumed entrance in the west and in the centre of the house was a hearth. Pottery from two different vessels was found inside the building. Five samples from the house were ^{14}C-dated and assigned the building to the period AD 100–600, i.e. the period to which most of the surrounding clearance cairns are also

Table 8. The numbers of ^{14}C-dated pieces of charcoal of hazel wood and charred hazel nuts distributed by context

Feature	Hazel wood	Hazel nut
Clearance cairn	9	1
Hearth	1	4
Pit	1	2
Post-hole		6
Stone setting	2	1
	13	14

dated. There were no long-houses found in association with the outbuilding, so the exact context of the house is unknown (Torstensdotter Åhlin et al. 2003: 17–21).

In the post-holes associated with the outbuilding 164 fragments of hazel nuts were documented, 105 of which come from the one and same post-hole. This is by far the highest concentration of hazel nuts in any context in this or the neighbouring sites; otherwise they occur as single pieces or in numbers up to six. In the post-holes some *Cerealia* of wheat and barley also occurred together with macrofossils indicating meadows and grazing lands (Regnell 2002, 2003b).

The character of the building with the central hearth indicates that this was primarily not a barn but a place where food was processed. The large amount of burnt hazelnuts seems to demonstrate that the nuts were roasted inside the building. From this we might conclude that the hazel bushes growing on the plots were an integrated part of agriculture, i.e. we are dealing with an agroforestry system.

The record from the Late Iron Age, AD 550–800, is presented in table 9. The use of trees in connection with agriculture seems to have changed. Instead of birch and hazel, oak and birch now dominate among the species-identified charcoal pieces. As a result of long-term grazing pressure and cultivation, hazel probably became less common in the landscape, as the soil was depleted of nutrients (Lagerås 2000).

Table 9. Species-identified charcoal from the Late Iron Age (1505–1255 BP) distributed by context

	Clearance cairn	Settlement	Grave	No.
Oak	5		2	7
Birch	4			4
Hazel	1			1
Aspen	1			1
Lime	1			1
	12		2	14

THE STRUCTURED USE OF TREES

The distribution of species-identified charcoal from archaeological contexts reveals a structured pattern. Certain kinds of trees occur in some periods but are rare or absent in other periods. There are also interesting contrasts between the two regions discussed and between contexts such as settlements versus fields.

The contrasting histories revealed by the record speak in favour of a structured geographical pattern that could not only be explained by references to a landscape gradually changing over time. These patterns are probably due to a combination of different landscape histories and different management practices.

As in pre-conquest North America, some of the patterns discussed above might be the outcome of man deliberately altering the composition of the wood by the use of fire. By using controlled fires the production of certain kinds of trees could be organized and structured.

The clearest indications of such a system are from the Western Småland Uplands where charred fragments of hazel wood were found below the clearance cairn providing evidence for hazel growing at the plots and testifying that the growth of hazel was an integral part of the agricultural system.

Since the Neolithic hazel has been the normal tree for wattle work such as hurdles, wattle and daub and woven fences; it was also regularly used as thatching wood. The specific advantage of hazel is that it can be twisted to separate the fibres and then bent at a sharp angle without breaking. Hazel wood can also be split to compensate for difference in thickness. For these purposes hazel is best coppiced with short rotation; after twelve to fifteen years' growth it becomes difficult to work (Rackham 2003: 206). Apart from the wood, the nuts are important as a source of food; these were collected on a yearly basis in the autumn.

During the Bronze Age birch dominates among charcoal found below clearance cairns in the Central Småland Uplands. Birch also occurs below clearance cairns in the Western Småland Uplands but in this region birch is not evident at the settlements. It seems as if birch in certain respects replaced hazel in the Central Småland Uplands when it comes to domestic use. Birch was used as fuel at the settlements, and presumably the birch was coppiced at regular intervals. When cut back at regular intervals the birches got multiple stems and the production of stakes, poles, firewood and wood for carpentry increased (Hæggström 1998: 34). Leftovers such as twigs and branches could be used as animal fodder.

If these interpretations are correct, it implies that the quite different life histories of cereals and trees were interwoven; this will be discussed below.

THE LIFE CYCLE OF TREES AND CEREALS

The species-identified charcoal found below clearance cairns is associated with agriculture and the growing of cereals. However, there is a substantial difference between cereals and trees when it comes to the notion of time. The production of cereals takes place within a one-year cycle where the starting and finishing points are determined by the sowing of the corn in the spring and the harvesting of the crop in the autumn. In contrast, the growing of trees operates within a very different time framework.

Trees like ash, elm and lime are managed to become very old since the purpose is the annual production of leaves taken from standing trees. In historical times these trees were usually pollarded at intervals of three to five years, although shorter intervals of one to two years may occur (Hæggström 1998: 34). The purpose of pollarding was to collect leaves and twigs as fodder for the animals.

In contrast, the ages of trees like birch and hazel were often limited by man to facilitate the production of fodder, fuel and sticks – i.e. these trees were coppiced. Written sources demonstrate that coppice rotation in eastern England during the Middle Ages was usually short or very short. The underwood was coppiced every four to eight years. After the Middle Ages coppice cycles usually became longer, ranging from ten to twenty years (Rackham 2003: 140).

Could it be that the very different cycles of cereals and coppiced trees were integrated at fields in the Bronze Age and Iron Age – i.e. that the yearly production of cereals was interwoven with the production of fodder, fuel and sticks from birch and hazel? Such a conclusion might help to explain the very large areas of clearance cairns dating to the Bronze Age and the Iron Age in the Småland Uplands.

Ever since the discovery of these fossil arable fields in the Småland Uplands in the 1980s their very large extent has been noted and discussed. There is agreement that the large extent is caused by the mobility of the plots, although the timing and rhythm of these movements and the causes behind them are disputed (Gren 1989, 1996; Norman 1989; Tollin 198; Widgren 1997; Lagerås 2000a; Lagerås and Bartholin 2003; Widgren 2003; Skoglund 2005).

A model proposed by Leif Gren emphasizes the combination of cereal production and the clearing of land from trees. As an effect of the reduction of tree cover, nitrogen was released in the ground, which facilitated the growth of cereals. Since this effect was short-lived, the plots had to be moved each year. Short periods of crop growing of about one year were followed by a longer fallow period of 20–40 years when the trees regenerated. Meanwhile, the trees growing on the former plots were coppiced and used for the production of timber and fodder. These quite long periods of fallow are based on analogies with 19th century Central Europe. In this model it is the lack of nitrogen that caused the abandonment of older agricultural plots and the establishment of new plots on a yearly basis (Gren 1989, 1996).

In a model proposed by Per Lagerås, the individual plots were supplied by manure and harvested regularly for a longer period. This model is based on analysis of a larger amount of ^{14}C-dated clearance cairns, interpretations of written sources (Tacitus) and analogues with archaeologically well-documented areas in southernmost Sweden. In contrast to the model proposed by Gren it is presumed that the plots were supplied by manure whereby they could be harvested for a longer period. However, at some time the plots were moved. It is suggested that the movement of plots was related to the movements of houses, which are presumed to have taken place at intervals of 20–30 years. The relation between cereal production and trees is not discussed in this model (Widgren 1997; Lagerås 2000a).

From the studies of species-identified charcoal beneath clearance cairns a third model can be proposed, implying that the movement of the plots was caused by the production of trees with certain qualities. If the coppicing of birch and hazel was carried out at intervals of four to eight years, as is known from medieval eastern England, the sowing of cereals could have taken place at a specific plot every four to eight years immediately after the hazel and birch had been cut down.

This model has many similarities to Gren's model, since the individual plots were cultivated only one year and then left for the coppiced wood to grow again. However, in other ways it resembles Lagerås's model, since there was a kind of stability in the system with a recurrent use of the very same plots in cycles of four to eight years, which presupposes that manure was spread on the plots.

One might wonder how the growing of trees could be combined with the growing of birch and hazel – even though the trees were cut down, tree stumps were left on the ground. The answer to this is that trees certainly grew in the clearance cairns which are regularly spread with a distance of 5 to 10 metres in between. The trees stumps would thus cause no extra impediment. Such a system would help to explain the predominance of charred remains of birch and hazel below the clearance cairns. In the western Småland Uplands hazel was an integral part of cereal production and the coppicing of hazel structured the rotation cycle. In the Central Småland Uplands birch was an integral part of cereal production and the coppicing of birch structured the rotation cycle. The background to the different use of trees should be sought in the different landscape histories of the two regions.

LANDSCAPE HISTORY AND THE MANAGEMENT OF TREES

The introduction of stone clearing varies from region to region, and three phases of expansion can be identified: c. 1500 BC, c. 800 BC and c. 200 AD. The very beginning of stone clearing around 1500 BC seems to be restricted to the central areas of the Småland Uplands (Skoglund 2005), while during the expansion phases around 800 BC and AD 200 more peripheral areas were also colonized and cultivated (Lagerås 2000a, 2002).

Around 1500 BC larger monuments, i.e. burial cairns and more rarely mounds, with impressive sizes ranging from 10 to 25 metres in diameter, replaced the gallery graves in the Småland Uplands. Often gallery graves were enlarged and modified into larger cairns, but cairns were also built in new locations in the landscape (Nilsson and Skoglund 2000). In parallel to larger cairns, there were also stone settings. Excavations have also proved the existence of grave-fields dating to the Bronze Age consisting of smaller graves built of stones, with round, oval shaped and rectangular forms that are not visible above ground (Åstrand 2009). Cup-marks and rock-carvings were also a part of this Bronze Age landscape, and while cup-mark sites are quite evenly distributed, rock-art sites have a more restricted distribution (Skoglund 2005, 2006, 2007).

This Bronze Age cultural landscape stands out very clearly in the Central Småland Uplands, which are rich in archaeological remains all the way back to the Mesolithic, but the Bronze Age cultural landscape stands out prominently, with the highest concentration of burial cairns exceeding 10 metres in diameter in the whole of Sweden (Figure 25) (Hyenstrand 1984; Gren 1996).

Figure 25. A monumental burial cairn, Tjureda Parish, Central Småland Uplands. Photo: Peter Skoglund

The situation in the Western Småland Uplands is different; here the Bronze Age cultural landscape is weak in its contours. Only about ten cairns are known from this area and among these the largest one measures 12 metres in diameter (Högrell 2000: 87). There is thus no sure evidence of Bronze Age settlements in the area based on visible monuments. However, archaeological excavations demonstrate that gallery graves in the river valleys just outside the study area were also used for burials in the Bronze Age. From the archaeological record it seems reasonable to conclude that colonization of the drumlins west of the river Lagan – in terms of permanent settlements – was an Iron Age phenomenon.

People living in the Western and Central Småland Uplands during the Iron Age thus inhabited two very different kinds of cultural landscapes. In the Central Småland Uplands there had been ongoing clearing of land and building of monuments since around 1500 BC, while in the Western Småland Uplands permanent settlements in the studied area was an Iron Age phenomenon.

This is reflected in the number of clearance cairns that can be attributed to different periods in the two regions discussed: in the Western Småland Uplands the average number of dated clearance cairns fluctuates from period to period (table 10), while the situation in the Central Småland Uplands is stable, with the amount of dated cairns remaining similar from the Late Bronze Age to the Late Iron Age (table 11).

Table 10. Number of clearance cairns in different periods and the average number of clearance cairns per 100 years. Based on tables 6, 7 and 9

Period	No. of clearance cairns	Average per 100 years
1400 BC–100 AD	14	0.9
100–550 AD	28	6.2
550–800 AD	12	4.8

Table 11. Number of clearance cairns in different periods and the average number of clearance cairns per 100 years. Based on tables 2, 3 and 5

Period	No. of clearance cairns	Average per 100 years
1700–500 BC	19	1.6
500 BC–400 AD	15	1.7
400–1000 AD	10	1.7

As a result of these different landscape biographies there are differences in the character of the clearing taking place in the two regions. In the Central Småland Uplands the stone clearing probably started around 1500 BC. Since there was a degree of mobility in the system, and probably also a gradual increase in population, more and more land was taken into consideration for agriculture. The stone-cleared area thereby expanded gradually towards lower altitudes, enlarging the land affected by human-induced fires and stone clearing. Around 500 BC the record indicates that areas close to alder swamps were used for agriculture. Moreover, during the period c. 750–500 BC *Calluna* heathland was formed on higher ground.

The use of areas close to alder swamps for agriculture and the formation of *Calluna* heathland at the transition between the Bronze Age and the Iron Age might be the result of an agricultural system under pressure. This situation should be understood from a historical perspective as agriculture involving the clearing of fields from stones at this time had gone on for about 1000 years in the Central Småland Uplands.

The situation is very different in the Western Småland Uplands. From around AD 100 a new cultural landscape was formed in an area lacking Bronze Age monuments. From this time onwards large areas were cleared for agriculture. Cereal production went hand in hand with the growing of hazel bushes. The hazel bushes were utilized for the production of hazel nuts and hazel woods. There are no indications from the Western Småland Uplands that the low-lying areas close to the alder swamps were used for cultivation.

These differences should be understood against the background of very different landscape histories. The landscape in the Central Småland Uplands had been cleared of stone continuously for agriculture since the Early Bronze Age while the Western Småland Uplands was only used non-intensively – primarily as grazing land – during the Bronze Age. The different landscape histories formed diverse conditions for the people living in the two areas around the birth of Christ. In the Western Småland Uplands leftovers of hazel from clearing the fields were left and burnt at the spot, while in the Central Småland Uplands, the presumably fewer hazel bushes were not integrated into the cultivation system. Instead heathland for grazing was formed by regular burning and low-lying, and presumably less suitable, areas close to alder swamps were cultivated.

Trees and the use of trees should therefore be studied in relation to the surrounding landscape. The utilization of trees involves as combination of traditions and existing natural resources. Moreover the natural resources are partially structured through people's practices. While doing things in the landscape such as building houses, clearing land, feeding cattle and constructing monuments, humans alter the relationship between grazing land and woodland and the composition and shape of trees.

Consequently, by their very presence in the landscape, people change the basis for their material culture. When landscapes are fully colonized and exploited, this will affect where and how to cultivate the land and graze the cattle, what trees can be used for fuels and timber and how the overall landscape is organized and perceived.

TREES AND RITUALS

There is a clear tendency for charred remains of birch and hazel to occur below clearance cairns in high numbers. This is in contrast to trees like ash, elm and lime that were rarely affected by fire. The latter phenomenon could partly be a result of the selection process when sending samples to the laboratory to be ^{14}C-dated. However, from a cultural-historical point of view we know that in northern Europe these trees were valued highly since they produced foliage of good quality, excellent for fodder.

This differences between birch and hazel and on the one hand and ash, elm and lime on the other hand might go back to the Bronze Age. There are many signs in the material hinting that these two groups of trees had a different social and ritual significance.

Hazel and birch were low-growing trees that were removed in connection with the clearing of land for agriculture. These kinds of clearances were labour-intensive and probably involved a wide range of people. Certain activities such as the felling of trees might have been carried out by people of a certain age or sex, while other acts such as the removal of stones and the collection of leaf fodder, fuel and nuts could have been done by women, men, children and elderly people.

This is different from the trees depicted on rock-art images discussed in the previous chapter. Through analysis of the pictures it was found that these trees represented high trees, probably ash, elm and lime, with a height of 20–30 metres. The climbing of these trees were probably done by people with a certain status, as indicated by the horned helmets worn by people in connection with these kinds of trees.

The collection of leaves from these high standard trees was ritualized. This is not to say that all collection of fodder from ash, elm and lime was ritualized. Probably the very first collection of leaves from these kinds of trees each year was turned into a ritual. Trees like the ash and the elm were also used in creation myths to symbolize the gender of woman and men.

The position of birch and hazel seem to have been different. These trees are almost totally absent from rock art. However, sticks from hazel are found in Scandinavian graves right from the Bronze Age to the Early Middle Ages (Andrén 2004). One reason for this might be that a concept of regeneration was attached to the fast-growing hazel which has an exceptional way of regenerating after being cut down. These kinds of notions are well known from historical folklore.

Based on these observations several oppositions between birch and hazel on the one hand and ash, elm and lime on the other can be identified, as summarized in table 12.

In conclusion, it seems as if different kinds of trees were not only treated differently but they also had different positions in rituals and mythology. The two different groups of trees have been discussed in Chapters 2 and 3 respectively. In Chapter 4 a third dimension will be discussed: the production, use and reuse of timber from oaks.

Table 12. Schematic overview of the social and ritual contexts of various trees

	Ash, elm and lime	Hazel and birch
Treatment	Selected and managed	Treated as staple products
Fire regimes	Unaffected by human-induced fire regimes	Affected by human-induced fire regimes
Age	Grow very old	Age of individual trees is decided by man
Fodder and timber collection	Require special knowledge and skills	General labour carried out by a broad range of people
Position in rituals	Mythology	Folklore

Chapter 4.
THE USE AND REUSE OF TIMBER

Large-scale excavations of long-houses – i.e. rectangular houses with two rows of inner roof-supporting posts that are typical of northern France, Holland, northern Germany and Scandinavia, have taken place for about 50 years. In the 1960s Carl Johan Becker conducted large-scale excavations of long-houses dating to the Iron Age in Denmark, and during the 1970s there was a breakthrough for the excavation of long-houses in Sweden. Before that houses were only known from those restricted areas where house remains of stones were visible in the landscape (Säfvestad 1995).

The identification of long-houses in the archaeological record is firmly connected to the technical and social developments in modern welfare society. The use of excavators in archaeology, combined with regulations requiring land with ancient remains to be excavated before development, have resulted in large areas with archaeologically excavated house remains in many urban areas of northern Europe (Borna Ahlkvist 2002; Björhem and Magnusson Staaf 2006; Arnoldussen 2008).

In many ways these results have profoundly changed our view of the everyday life of prehistoric man. Because of the long-houses we have obtained a framework for understanding social and economic dynamics from a household perspective and our comprehension of the prehistoric landscape has been enriched.

However, when it comes to our perceptive of the houses *per se*, archaeologists tend to approach these as if they were absolute types fixed in space and time. Since they are regarded as types they are ordered in chronological series, with certain traits in the construction deciding the age of the individual house. This approach does have benefits, of course, and it has laid a foundation for our understanding of prehistoric houses. It works well on a general level since architecture changes as a result of social, economic and technical changes in society. Houses are a good reflection of the way social relationships are organized and how people make their living out of the land – when these circumstances change houses change too (Björhem and Skoglund 2009).

However, the typological approach misses one important dimension: houses are by character building kits. They are composed of various pieces of wood designed to fit into each other and held together by wooden nails and ropes. This kind of construction could easily be deconstructed and split into pieces and put together at a new place. If timber was reused the timber itself laid constraints upon the construction of a new house. In these cases houses were not designed from scratch but modified from a pre-existing plan.

Working at site level, there is an obvious risk that certain traits in the building which result from ad hoc solutions because of timber reuse are regarded as typological traits with chronological significance. Therefore one may wonder why archaeologists rarely address the question of timber reuse.

A major factor is probably that archaeologists tend to focus on the negative factors of timber decay. Posts dug into the ground have a limited lifetime. Depending on the type of species chosen and how the timber is treated, the durability varies. There is a broad range of literature on this, and the estimates range from a decade to a century, with 20–40 years as an often cited number (Arnoldussen 2008: 88 with references).

If one or a few of the roof-bearing posts rotted at ground level and sank into the ground, this was probably not a great problem. Because of the elasticity in the construction the tensions caused by a single shrinking post were levelled out to other parts of the building. However, when several posts had rotted substantially it was necessary to build a new house, since it was very complicated or even impossible to replace several posts with new ones when the house was standing. There is nothing to suggest that the rest of the building was also rotten or in need of replacement at this point of time, since different kinds of wood in different parts of a building have their own specific permanence (Figure 26).

If the house was not burnt down, or the timber was not affected by noxious insects, the greatest threat to the

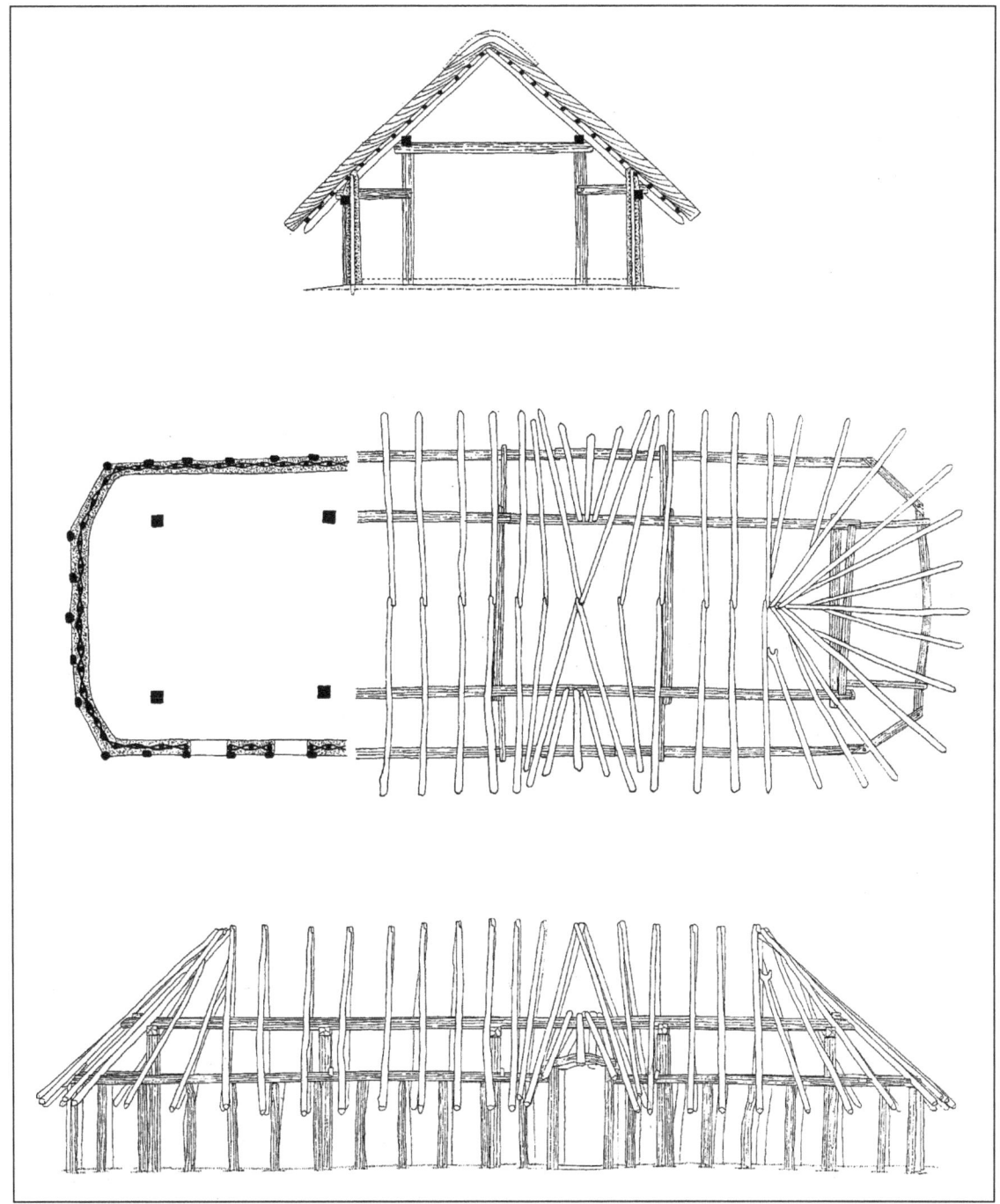

Figure 26. Plan and section of a reconstructed Iron Age long-house at Lejre, Denmark. Note that substantial parts of the timber construction are not dug into the ground, which would make them last longer than the roof-bearing posts and the wall posts. From Draiby 1991: 117.
Drawing: Bente Draiby (Lejre Forsøgscenter)

timber was probably damp from the ground, rain and snow falling on the building, and the damp caused by cattle if these were stalled inside the building. These threats will affect different parts of the building in various ways. Stalls will probably start to rot earlier than a dwelling area or a storeroom, and the vertical posts dug into the ground will rot earlier than the horizontal beams in the upper part of the construction. However, a prerequisite for the long preservation of the horizontal beams is that they are protected by a watertight outer roof.

If timber is kept dry and is not threatened by fire or noxious insects, it can last a long time. The breakthrough of dendrochronology as a method has not only changed previous views about the dating of certain archaeological materials from excavations, but has also brought to our attention a heritage of still used timber constructions with an unexpected and impressive age.

From dendrochronology we know today that in northern Europe there are timber structures that are very old. For example in England timber-framed aisled halls have been dated to late 12th and early 13th centuries. These houses varied in function and status. Among these are both smaller buildings and buildings of higher status such as manor houses, or their equivalent, and a bishop's palace (Walker 1999).

Another example is the occurrence of a large number of medieval timber houses in central and northern Sweden that have survived into our days. Through dendrochronology 134 timber buildings in the province of Dalarna alone have been dated to the period 1285–1600 and amazingly 35 of these houses were built before the year 1350. The original use of these houses was often as stores or threshing barns, although the function of the building often changed as time passed, and as a result of this many of them were moved to new positions in the landscape (Raihle and Homman 2005).

On a quantitative level, however, the medieval secular timber buildings are probably outnumbered by medieval church roof trusses. In northern France and Belgium there are preserved church trusses from the 11th century. From Norway there are 27 known church trusses from the 12th century and in Sweden there are 43 known church trusses from the same century. In England there are a few examples of trusses from the 13th century. Adding the trusses from the succeeding centuries, it is clear that there is a very large number of very old timber roof constructions in northern Europe (Linscott 2007).

These examples, of course, cannot be directly translated to a prehistoric situation, but they nevertheless demonstrate that there is nothing regular in the way a timber structure decomposes. The decomposition or preservation of a timber construction is dependent upon many various factors such as the environment, the building technique, the function of the house and the house-building tradition in a wider sense.

From these observations it seems reasonable to conclude that various parts of the timber inside a prehistoric longhouse could last for very different periods of time. Therefore it seems important to consider the life history of the individual timber in order to gain a better understanding of prehistoric long-houses. In the next section there will be an attempt to discuss timber use from this perspective.

THE LIFE CYCLE OF WOOD AND TIMBER

Prehistoric houses with preserved timber only occur in rare instances. Well-known examples are the Swiss lake settlements, the Irish crannogs and certain sites along the coast of the North Sea in Germany and Holland.

From a quantitative perspective the most revealing finds informing us about ancient wood crafts are probably ships, wagons and ploughs. Of these the wagons are of particular interest since they are complicated constructions linked to ordinary land-based life and thereby occur in wide geographical setting. The prehistoric wagon finds from north-west Europe have been catalogued and analysed by the Danish archaeologist Per Ole Schovsbo (1987).

Based on the information from 560 pieces of wagons occurring at 200 different localities Schovsbo is able to draw important conclusions on the development of the wagon in northern Europe from the Neolithic to the 11th century AD. During his studies, which took him to a lot of museum stores, he also acquired a picture of the use of wood in prehistory.

An important aspect of the use of wood in wagons was that the trees after felling were stored in wet environments to reduce the tensions in the wood. By storing the wood in waters or bogs for a period of up to several years, the softness of the wood maintained though the tensions in the wood were reduced; some of these qualities were maintained even after drying, since the process helped the wood to resist variations in humidity in the atmosphere without splitting or becoming deformed (Schovsbo 1987: 49–50).

In support of the existence of this wet wood technology, apart from information drawn from the technical construction of wagons, there is also the archaeological record and the find circumstances. Out of the 200 find locations catalogued by Schovsbo, as many as 20 revealed preparatory work. A majority – or 15 of these – are from bogs or presumably wet environments. Many of these workpieces were certainly stored in wet environments in order to reduce the tension in the wood, but for some reason they were never resumed (Schovsbo 1987: 54).

In Scandinavia the wet wood technology seems to dominate during prehistory even though hard wood technology also occurs. This situation changed in the Middle Ages, but the technique prevailed in some typical rural handicrafts such as the making of clogs and baskets (Schovsbo 1987: 49).

From his study of wagons Schovsbo suggested a model to illustrate the preparation and use of timber in wagon constructions. He identifies six different stages in the process of making, maintaining and reusing a wooden wagon (1987: 50):

1) The choice of timber and the felling of the tree.

2) The removal of the bark and branches, the selection of different pieces of wood for different constructions, transportation of the material to the place where it will be stored.

3) The preparation of workpieces close to the storage place.

4) The actual making of specific constructional details with certain qualities.

5) The maintenance and repair of broken wooden details, the replacement of worn-out pieces with new constructional details of wood.

6) The discarding or reuse of wooden material from wagons that had gone out of use.

Inspired by the work of Schovsbo, a similar sequence of events will now be sketched concerning house timber. Because of a general lack of preserved timber from prehistoric buildings this is of course a difficult task. However, in the following an attempt will be made to discuss the life history of house timber under the headings: *The choice of timber, Storage and preparatory work, The making of constructional details, The maintenance and repair of wood* and *The reuse of timber*.

The choice of timber

There was probably a major difference between the making of small structures such as ards and ploughs on the one hand and buildings on the other. Substantial parts of wagons and ploughs were made out of wood with specific shapes. This is true for the crock ard that was made out of one large piece of wood where the only thing added was the handle. For example, the crock ard from Vebbestrup in northern Denmark was made out of birch, where the stem of the birch together with one of the branches was used to get the right shape. To this only a handle of hazel was added and a piece of oak to connects the plough to the oxen (cf. Glob 1951: 16–19).

Different trees had different qualities which made them more or less suitable for a specific function. Therefore the carpenter in these examples chooses the wood pieces carefully. The ard was made out of a solid piece of wood that needed to be bent at the right angle if the ard was to be able to fulfil its purpose. The occurrence of trees with specific shapes was something that humans could only affect very little.

In contrast to these examples the house-building process required not small amounts of wood with specific shapes but large amounts of straight timber. Here the focus was on getting a large quantity of straight timber with various diameters that could fulfil different purposes in the house construction.

The best way of solving this was not to rely on nature alone but to modify the woodland in order to obtain larger stands with straight trees of similar sizes. This situation would occur if the woodland was coppiced on a regular basis. Cutting back the trees made new trees grew up from the tree stump. If this was done in an ordered way, different places in the landscape would supply trees of a certain age and diameter.

In those rare cases when it is possible in an archaeological context to study large amounts of preserved wood, such as preserved Mesolithic fish traps made of hazel, it sometimes turns out the age of the individual pieces of timbers is not random but clusters in age classes. An interpretation of this phenomenon is that the wood was taken from woodland that was coppiced at regular intervals (cf. Bartholin 1996).

Preparatory work

As noted above, tensions caused by movements in the wood can be distributed to other parts of the building without affecting the stability of the house. Therefore the storage of house timber and the reduction of tension in the wood was probably a less difficult process compared to wood used for the wagons discussed by Schovsbo.

In those cases timber was not available very close to the construction site; the timber was probably measured and cut up in pieces at the felling site, in order to facilitate transportation. This was done immediately after the felling when the timber was still soft and could be worked by axes.

When an Iron Age long-house was re-constructed at Lejre in Denmark, 20 oaks of an estimated age of 45 years were used. In addition to the 20 oaks younger oaks were also used for the roof structure. Moreover, willow and hazel were used for the roof and wall.

At the felling site the 15-metre long timbers were divided into top and root pieces of a length of about 7 metres each. These pieces of timber had different qualities and were later used in different parts of the building. The use – and partly also the length – of different parts of the tree was in this example decided at the felling site and not at the construction site (Figure 27) (Draiby 1991: 121).

The making of constructional details

At the Priorsløkke site in Denmark roof-bearing posts from house constructions dating to the third century AD were reused in a defensive wall. Some of the posts survived in wet environments and some of these were placed upside down. It was therefore possible to study the roof-bearing posts, which were made out of forked trees so that the upper beam could rest in a hollow (Kaul 1985: 178–181). On this occasion the different pieces of timber were hold together by the shape of the timber pieces themselves and probably rope or bast. The use of wooden nails in ancient long-houses was probably restricted since the drilling of holes required so much labour (Figure 28).

A similar technical solution is highlighted in the oldest preserved Scandinavian ship, the Hjortspring boat from c. 350 BC found in a bog in Denmark. The ship was built of planks from lime sewn together with bast of lime. Since no rivets were used, all details such as clamps for fastening the frames had to be carved out of the same plank (Kaul 1988; Crumlin Pedersen and Trakadas 2003; Larsson 2007: 80–82).

The situation changed in the Late Iron Age. In contrast to the Bronze Age and Early Iron Age we now know of boats made from radially split wood. This was due the technical developments during the course of the Iron Age. Improved axes facilitated the making of planks by the radial splitting technique (Larsson 2007: 85–91). Whether or not these changes were also reflected in the way houses were built we do not know, but it seems plausible.

THE SIGNIFICANCE OF TREES: AN ARCHAEOLOGICAL PERSPECTIVE

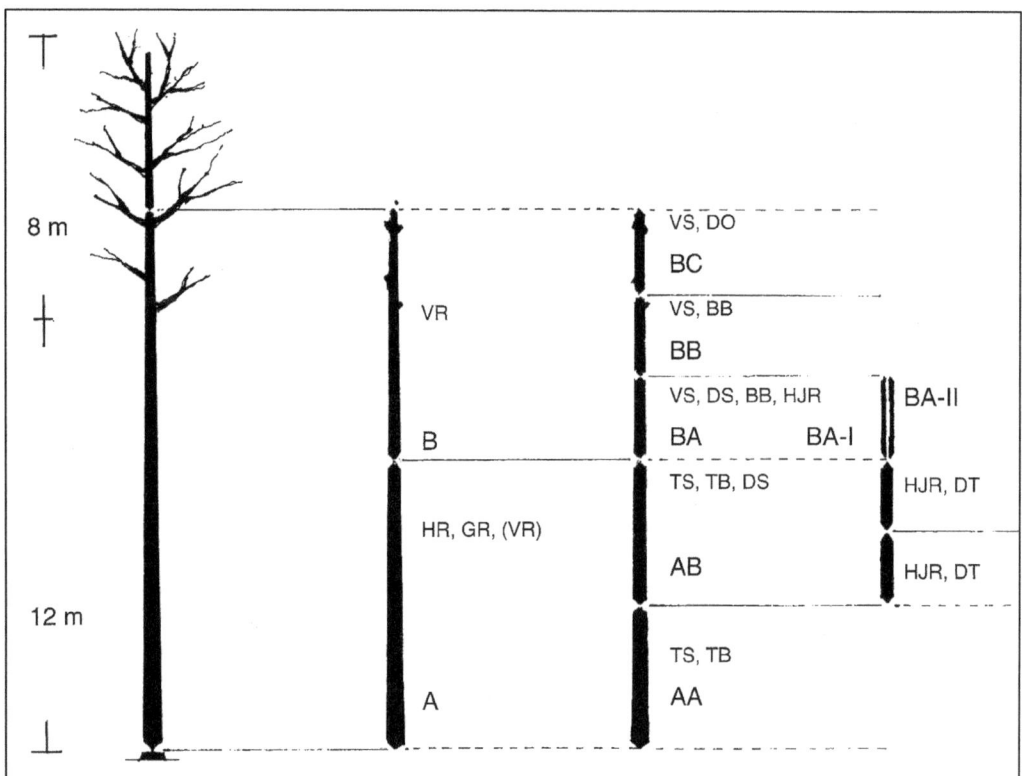

Figure 27. Diagram showing possible timber use from a tree. A=Lower trunk; B=Upper trunk; HR=High beam; GB=Gable head; VR=Wall head; TS=Roof-bearing posts; TB=Crossbeam; DS=Door posts; VS=Wall posts; BB=Connecting beam; HJR=corner head; DO=Door head; DT=Doorstep. From Draiby 1991: 121. Drawing: Bente Draiby (Lejre Forsøgscenter)

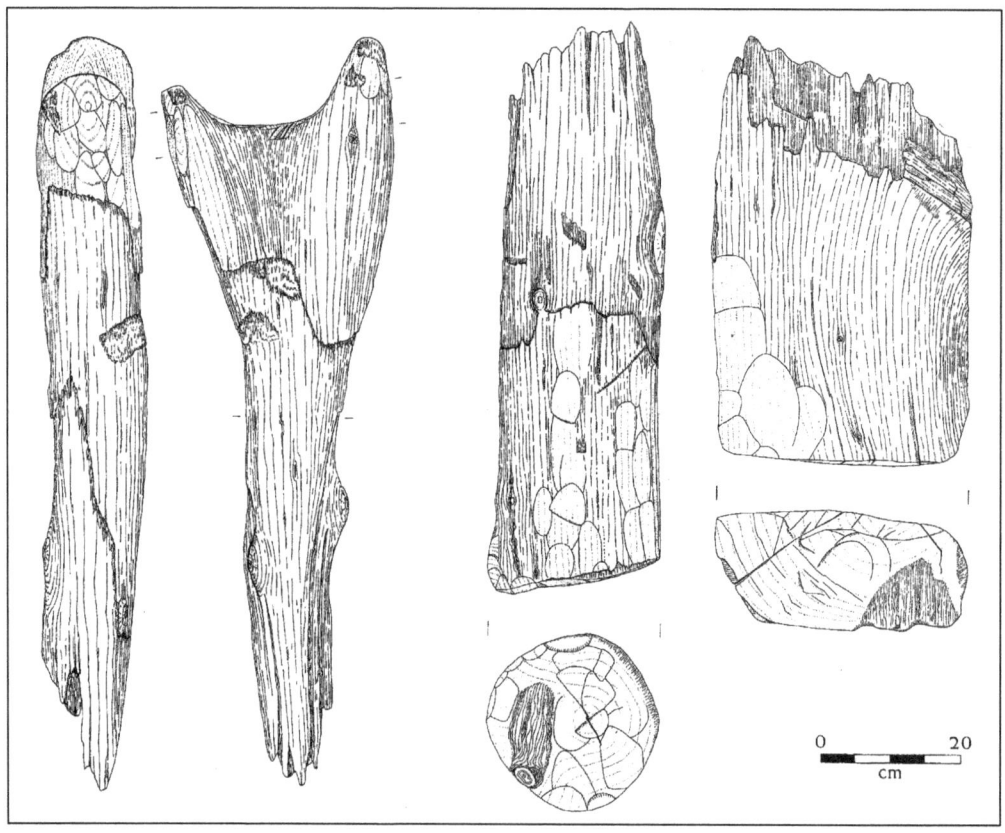

Figure 28. Wood from the Priorsløkke site at Horsens, Denmark. To the left is the upper part of a roof-bearing post; in the middle is the bottom of a roof-bearing post; and to the right is a plank, presumably a wall plank. From Kaul 1985: 181

Colourings with the shape of radially split wood have been recorded in Iron Age houses (Björhem and Säfvestad 1993: 227).

The maintenance and repair of wood

Replacing and repairing wooden details was probably a natural thing to do in any traditional society. This phenomenon can also be studied on the Hjortspring ship. When the boat was put in the bog it was old and had gone through several repairs. At a few places in the construction there are damaged parts measuring a metre. These were repaired by putting in new pieces of wood fastened with bast and sealed with a fixing paste (Kaul 1988: 20).

In the long-houses this process can be studied through the occurrence of additional roof-bearing posts that were probably not part of the original construction, or asymmetries in houses that might be the results of extensions. Both these phenomena occur regularly in houses, testifying to the maintenance of individual buildings (Arnoldussen 2008: 392).

The reuse of timber

The ancient building technique discussed above facilitated timber reuse since no, or very few, rivets were used and houses were therefore built like kits; this become obvious when studying the Priorsløkke site where the roof-bearing posts were reused in a defensive wall.

Moreover, in one of the post-holes in a smaller outbuilding at the Priorsløkke site a forked colouring in the post fill was identified. The interpretation in this particular case is that a roof-bearing post from a larger house that had rotted at ground level was reused. The damaged part was cut away and the post was put upside down in a smaller and lower building (Kaul 1985: 180).

Sometimes similar conclusions could be drawn from the analysis of house plans. For example, at a Late Neolithic settlement in Scania, south Sweden, it was noticed that the two house plans of similar type partly overlapped each other. In these kinds of houses the roof-bearing construction and the walls were supposed to be independently constructed. A possible explanation for the partial overlap of the house plans is that the roof structure belonging to the older house was still standing as the wall in the new house was built of material taken from the older house (Björhem and Säfvestad 1989: 74).

From these and other observations, two important conclusions can be drawn that are important if we are to understand timber reuse. Pieces of wood were individual in a way that they are not today. Wood with specific shapes and characteristics was chosen for certain constructional details. Therefore a piece of timber could not be randomly replaced but had to be replaced by a specific kind of timber. Because a lot of planning and effort was invested in getting the right kind of timber, the reuse of timber was probably an attractive alternative.

The second observation is that houses, boats and wagons functioned as kits that could be taken to pieces. Since no iron nails were used, pieces of wood were fastened together by clamps, bast or rope and in certain cases by mortise and tenon joints. The occurrence of the latest technique in a Bronze Age context is demonstrated by the preserved upright posts and cross-beams from the cult house at *Bargeroosterveld* in Drenthe, Holland (Brongers and Woltering 1978: 26–27; Harding 2000: 309). In summary, Bronze Age houses consisted of various pieces of wood which easily could be reused in new structures.

Is it possible to study the life cycle of timber from an archaeological perspective? Some of the issues outlined above are dependent upon actual preserved timbers, which are rarely found at excavations, while others, such as the repair and extension of standing buildings, are regularly discussed on the basis of house plans. Other phenomena in the chain discussed above, such as the choice and use of various kinds of timbers in buildings or the reuse of timbers, are rarely discussed from the archaeological material.

In the following I will first try to discuss the use of timber in an Iron Age house using data from dendrology, and then I will go on to discuss the possible reuse of timber by analysing Bronze Age house plans from houses succeeding each other on the same site.

READING WOOD INTO HOUSE PLANS

As part of the Öresund Fixed Link Project – which managed the construction of a bridge between Copenhagen in Denmark and Malmö in Sweden – large-scale archaeological excavations took place around the city of Malmö in Sweden (Figure 29) (Björhem and Magnusson Staaf 2006). The archaeological project was divided into different sub-projects addressing various topics. One of these sub-projects was labelled *The Functional Landscape* (Eliasson and Kishonti 2007).

As part of the programme a broad array of analyses were applied to a wide range of features and constructions. Among these was dendrology and species identification of charcoal. Charcoal from post-holes belonging to ten houses was identified. The most ambitious programme was applied to House III at Lockarp; from this house not only charcoal from the roof-bearing posts but also wall posts and posts from the inner construction were identified to species (Eliasson and Kishonti 2007: 162–188).

This building was 38.90 metres long and 5.20–5.70 metres wide. The outer wall consisted of smaller post-holes and in parts of the construction ditches. The gable in the west was rounded while the gable in the east was straight. Inside the house were 10 trestles situated at a distance of 1.70–4.60 metres. In connection with the house, and oriented roughly at right angles to it was an outbuilding measuring 6.0 x 3.5 metres that lacked trestles; the roof rested on the outer walls. The

Figure 29. Map showing the location of the areas discussed in the text

outbuilding – interpreted as a barn – was attached to a fence that encircled a rectangular area around the long-house (Figure 30) (Eliasson and Kishonti 2003: 36–43, 56–61).

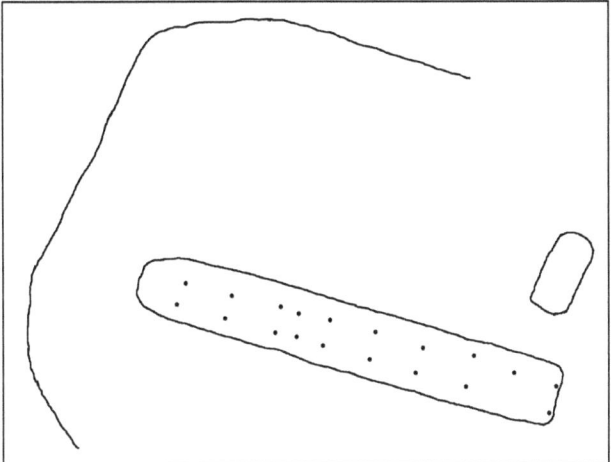

Figure 30. House III at Lockarp with surrounding fences and an outbuilding. From Eliasson and Kishonti 2003: 57

The long-house, the outbuilding and the fence have been dated with a total of seven ^{14}C samples to the third century AD. This farm is the latest in a series of three long-houses on the site. The infill in the post-hole and the macrofossil analysis demonstrates that the house was burnt down and thereafter the farm most probably was abandoned (Eliasson and Kishonti 2003: 36–43, 56–61). This is in line with recent research demonstrating a general relocation of the settled landscape in this part of Scandinavia around 300 AD (Björhem and Skoglund 2009).

Apart from species identification of charcoal, macrofossil analysis, phosphate analysis and MS analysis were carried out. Based on these results and supplemented by information from the archaeological record, the excavators have suggested a room division of the house that is summarized in table 13 (Figure 31). In the west was an entrance room followed by a kitchen and a living room; then came a barn, a threshing barn and a stable and finally in the east there was another kitchen (Eliasson and Kishonti 2003: 36–43, 56–61; Eliasson and Kishonti 2007: 208–212).

We thereby get a tripartite division of the house, with the major living area in the west, followed by rooms connected to the keeping of animals and finally a smaller living area in the east. This division suggested by Eliasson and Kishonti will be followed here with one minor modification. The western living section will be

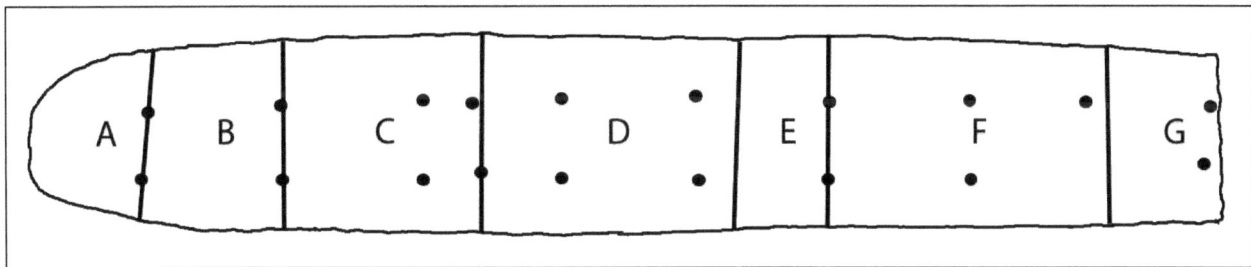

Figure 31. The inner division of the houses at Lockarp: A=Entrance room; B=Kitchen; C=Living room/Storeroom; D=Barn; E=Threshing barn; F=Stable; G=Kitchen. From Eliasson and Kishonti 2007: 210

Table 13. The room division at House III, Lockarp, Sweden as suggested by the archaeological record, Macrofossil analysis, phosphate analysis and MS analysis. Information from Eliasson and Kishonti 2003: 36–43, 56–61; Eliasson and Kishonti 2007: 208–212

Distance from west gable (m)	Archaeology	Analysis	Interpretation
0.0–4.0	Entrances in the north Entrance in the south		Entrance room
4.0–8.4	Concentration of finds	*Cerealia* concentration Phosphate concentration	Kitchen
8.4–13.0	Finds occur Entrance in the south		Living room Storeroom
13.0–23.6		Macrofossil concentration	Barn
23.6–26.2	Entrances in the north Entrance in the south	Macrofossil occur	Threshing barn
26.2–34.6	Entrance in the north	Remains of fodder?	Stable
34.6–38.9	Hearth		Kitchen

defined as 0–18 metres from the west, the middle section including barn and stable will be defined as 18–34 metres, while the eastern living section will be defined 34–38.8 metres from the west. This minor modification is due to the interpretation of the species-identified charcoal, as will be discussed below.

The function of the eastern room is not discussed in detail by the excavators, but they refer to an example at Gene in northern Sweden where a smaller kitchen occurred in connection with a stable (Ramqvist 1997). In this case the hearth in the room was supposed to heat the fodder for the animals during the winter. This is probably not the function of the kitchen at Lockarp, situated in southern Sweden with a warmer winter climate.

Instead we can suggest a division of the living area into two sections with a major living area in the west for the family in control of the house, and a minor living area in the east for servants, or people working on the farm. This interpretation is supported by a recent study underlining that the Iron Age houses in the area do not primarily reflect independent nuclear families but households differing in social status – probably there were both families that were in control of their own farms and people that lacked farms of their own (Welinder 2009: 212–215).

Because charcoal in post-holes belonging to the house was species-identified it is possible to compare the results of these analysis with the room division suggested above. In all charcoal from 30 post-holes were identified. Of these 15 were roof-bearing post-holes, 11 were outer wall post-holes and 4 were post-holes connected to the inner construction of the building. This means that 15 out of 20 roof-bearing posts, about half of the post-holes connected to the inner construction, and a smaller sample of the wall post-holes were examined (Eliasson and Kishonti 2003: 237–240, Plate 4; Eliasson and Kishonti 2007: 168–174).

The charcoal in the post-holes comes from three different sources, which complicate any interpretation of the material. It can be part of the construction that most probably was burnt down (see above), it may represent wooden artefacts or leaf fodder stored in the houses, and finally it may represent charcoal from the hearths inside the house that was swept down in the post-holes while the house was still standing.

The lack of macrofossil plants indicating hay fodder, and the lack or low occurrence of charcoal from trees well known for being used as fodder, such as ash, lime and maple, speaks in favour of the house either being burnt down during the summer or being cleaned out before it was burnt.

Oak totally dominates the sample, occurring in very high numbers, and beech occurs in quite large numbers, while hazel, hawthorn/apple/rowan and alder also are represented with some numbers and finally sweet cherry/bird cherry/blackthorn, willow together with ash occur in low numbers (table 14).

Table 14. The number of charcoal pieces from various species at House III, Lockarp, Sweden. Data from Eliasson and Kishonti 2007: 172–173

Species	No. of pieces
Oak	321
Beech	61
Hazel	22
Hawthorn/apple/rowan	21
Alder	15
Sweet cherry/bird cherry/blackthorn	9
Willow	9
Ash	3
Buckthorn	1

From the distribution of these species they can be divided in two groups: species with a clear concentration in the two living areas and species that occur in both the living areas and in the middle section of the house.

Species that are restricted to the living areas are hazel, hawthorn/apple/rowan, sweet cherry/bird cherry/blackthorn, alder maple and ash. Of this hazel, hawthorn/apple/rowan occur in the highest number and they are also well represented in features at the site that can be related to fire (Eliasson and Kishonti 2007: 164).

We may therefore presume that these pieces of charcoal originally were wood collected for fuel, used to fire the hearth, and thereby they became charcoal dust that was swept down during cleaning into the shallow holes created by the shrinking infill around the posts. The same probably also goes for the smaller amount of alder, maple and ash found in the living areas. However, it cannot be ruled out that some of the material discussed above originally had other functions such as building material or leftovers from leaf fodder with a secondary function as fuel (Regnell 2003a).

Oak and beech have a different distribution pattern. These species are represented both in the two living areas and in the middle section of the house. Oak was most probably used in the roof-bearing structure. Oak is by far the best-represented species in the house; it occurs in more than 300 fragments, which is much more than the other species that occur in less than c. 60 fragments each. The estimated age of the oak is also higher than that of the other species. In several of the post-holes there is oak with an estimated age of around 50 or 75 years, compared to an average estimated age of between 10 and 40 for most of the other species on the site. These observations clearly speak in favour or oak being used as timber for the roof-bearing posts, the wall posts and posts connected to the inner construction, as previously suggested by Laila Eliasson and Ingela Kishonti (2007: 169, 174).

What about the beech that has a similar distribution pattern to oak, even though it occurs in lower numbers?

Table 15. The distribution of oak with various ages at House III at Lockarp, Sweden. Data from Eliasson and Kishonti 2007: 172–173

Distance from west gable	0–18 metres			18–34 metres		34–38.8 metres	
Kind of feature	inner posts	wall posts	roof-bearing posts	wall posts	roof-bearing posts	wall posts	roof-bearing posts
Oak <75	2	3					
Oak <50	1		4	1	1		1
Oak <40							1
Oak <25			1	2	4	2	

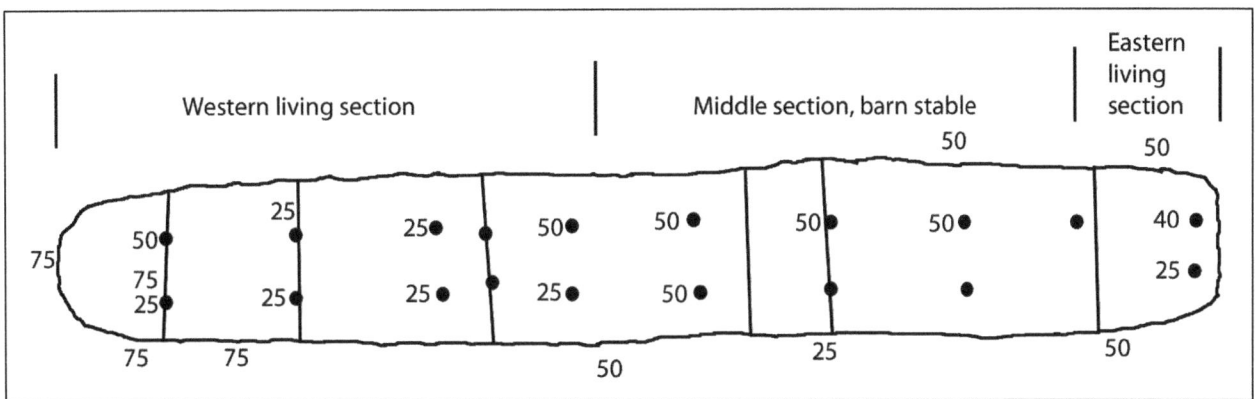

Figure 32. Plan showing the distribution of oak of different age, House III Lockarp. Room division according to Eliasson and Kishonti 2003 and above is the author's interpretation based on the charcoal evidence. Information from Eliasson and Kishonti 2003: 37; Eliasson and Kishonti 2007: 172–173

From historical sources we know that beech was rarely used in house constructions because it rots easily. However, since beech is a very solid wood it was used in certain structures such as threshing floors in barns. Another use of beech was for making furniture (Emanuelsson 2002: 76).

However, the major use of beech was for fuel, since it burns steadily with a bright light. At the Lockarp site beech is not very well represented in these contexts, occurring in one out of 12 features related to fire (Eliasson and Kishonti 2007: 164). A possible explanation for the occurrence of beech, apart from being used for furniture, could be that it was used as lighting sticks. Especially during the winter time there must have been a need for light not only in the living section of the house but also in the stable. This is one use of beech that would help to explain the distribution pattern of carbonized beech.

The charcoal was not only identified to species but the age was estimated as well. Both the species determination and the age estimation were made by Thomas Bartholin, with long experience of species identification. The age estimation was primarily done to facilitate the selection of charcoal with a low age for radiocarbon dating. This was subjective and was never indented to be used in an archaeological interpretation (Eliasson and Kishonti 2007: 193).

Despite this the estimates are used here together with the species identification to see whether there are any meaningful patterns in the house that can shed light on the construction. Judging by the species identification and age estimation there is a tripartite division of the house. The observations are summarized in table 15.

From the table it is clear that oak with different estimated ages is unevenly represented in the house. Oak with an estimated age of 50 years dominates in the roof-bearing post-holes in the western section, while oak with an estimated age of 25 years dominates in the middle section, which is interpreted as a barn and a stable. If the estimated ages are correct, the timbers used in the roof-bearing posts in the western living section were older and had a different character than those timbers used in the middle section. The situation in the eastern living section is less obvious but the age of the oak in the roof-bearing posts resembles more the western living area than the middle section of the house (Figure 32).

A possible explanation is that there was a heavier roof construction above the two living sections which demanded roof-bearing posts of a different quality or a larger diameter. Another interpretation focuses on the social significance of various choices of timber. At the Iron Age site of Hodde in Denmark the size of timber seems to have been used as a social marker since one of the houses in the village was constructed out of timber with a larger diameter than surrounding houses (Hvass 1985: 100–101).

The distribution of oak with an estimated age of 75 years is interesting since it has a clear concentration in postholes either in the outer wall or the western inner construction. Oak with an estimated age of 75 years found in the wall would indicate planks made out of larger pieces of timber. There are examples of both planks and smaller posts in walls from Iron Age houses in Scandinavia (Draiby 1989: 116). At Lockarp the age determination of the wood seems to indicate that the use of planks made of older wood was restricted to the western living section.

Yet another phenomenon is revealed by the age determination: the oak cluster in age intervals of 25, 50 and 75 years (table 16).

Table 16. The distribution of charcoal of oak with various ages from House III, Lockarp Sweden. Data from Eliasson and Kishonti 2007: 172–173

Estimated age	No. of samples
<25	9
<40	1
<50	8
<75	5

The occurrence of timber with specific and different ages should not be taken too literally since we are dealing with estimates of age and not ages measured in exact years. However, the distribution of the wood in three different age classes is something one would expect if the trees were taken from a coppiced wood.

In coppiced wood various parts of the wood were cut back at regular intervals. Therefore it is possible that the building material for the roof-bearing posts, the wall posts and the inner construction of the house was taken from coppiced trees with an age between about 25 and about 75 years. Coppiced woodland management facilitated the accessibility of trees with different qualities that could be used in different parts of a house construction (Eliasson and Kishonti 2007: 193).

In summarizing it seems as if the different functions and statuses of the house were reflected in different choices of timber for different parts of the house. In the stable and barn the roof-bearing posts were of younger age than the living sections. The older and presumably larger roof-bearing posts in the western and eastern living sections might indicate a loft over these parts of the building. Finally, the use of the oldest timber was limited to the western living section where it appears in the outer wall and inner construction. An explanation put forward is that this reflects that the use of planks made out of larger pieces of was restricted to the western living section. This part of the house is also presumed to have had a higher status than the eastern living section. The species identification hints at the timbers having been taken from a coppiced wood that was cut back at intervals of about 25 years.

The Lockarp house seems to reveal that houses were built of various kinds of timber with different qualities and that wood management stretching over generations proceeded the actual building of the house. At the Lockarp site the timber was burnt down and reuse was not an option. But what about those cases when houses were not burnt down – could it happen that timbers were reused in other houses? This leads us to the next issue – tracing the possible reuse of timber by analysing house plans.

THE REUSE OF TIMBER AS INDICATED BY HOUSE PLANS

In this section the possible reuse of timber will be examined. The inspiration will be taken from three Bronze Age settlement studies which have all been part of PhD projects. We will start by getting an overall impression of the durability and character of Bronze Age settlements from Stijn Arnoldussen's study of Dutch settlements (2008); then we will turn to central Sweden to consider Hélène Borna Ahlkvist's ideas of formalized architectural traditions along separate family lines (2002). Finally we will turn to southernmost Sweden and address the topic of neighbouring and very similar house remains previously discussed by Sten Tesch (1993) and Anna Gröhn (2004).

The running theme of the discussion is the possibilities of tracing timber reuse, which will be elaborated in the final section.

The Dutch central river area

In an extensive and comprehensive study Stijn Arnoldussen has studied the Bronze Age house remains of the central Dutch river area. From our point of view this study is interesting because it underlines the continuous use both of the individual houses and of the house site.

The major factor deciding the lifetime of the house was the durability of the posts dug into the ground that gradually decayed at ground level once the house was erected. Traditionally the lifetime of the roof-bearing posts has been regarded as limited and not exceeding 50 years.

However, Arnoldussen has been able to analyse a few houses from where there are both ^{14}C dates and dendrochronologically dated timber. In these cases it seems clear that the houses were repaired and new outbuildings were added to already existing houses over a substantial period of time. From these data the lifetime of a long-house could be estimated at more than 50 years, perhaps approaching a century (Arnoldussen 2008: 90–92). Similar results are also reported for Bronze Age houses from southern Sweden (Björhem and Säfvestad 1993: 141).

Because of the long use of the houses they were commonly repaired, and in about one third of the Middle Bronze Age B houses there is evidence of repair. About one out of ten houses was extended during its lifetime, probably as a result of new needs arising as the composition and status of the household changed (Arnoldussen 2008: 392).

From Arnoldussen's extensive data it is also clear that once a particular house location was decided, new houses were often built close to the older and abandoned houses. A house site thus maintained its function for a long time, ranging between 50 and 300 years.

A particular case is when houses were rebuilt on the same spot. The term rebuilding refers to those cases where two or more superimposed plans of structures are found on top of each other and having the same function, dimension and orientation. These similarities indicate that the houses were erected by the same group of people (Arnoldussen 2008: 73–74).

In four cases in Arnoldussen's study area it has been possible to demonstrate striking similarities between houses that were rebuilt on the same spot. A very illustrative example is from *De Bogen*, where a house was rebuilt four times with a more or less identical roof-bearing structure (Arnoldussen 2008: 392–393).

The houses from *De Bogen* raises the question what these similarities represent. Are they merely the outcome of a tradition or do they also tell us something about the use of timber in the Bronze Age? This question will be discussed in more detail against the background of the evidence from Pryssgården in eastern central Sweden.

Pryssgården in eastern central Sweden

One of the major Bronze Age settlement excavations conducted in Sweden was undertaken outside the town of Norrköping at a place called *Pryssgården* in eastern middle Sweden during the years 1993–94. The excavated area measured 80–90 x 800 metres. At this site 21 long-houses and 12 smaller houses dating to the Bronze Age came to light. The houses are dated to the time span 1200–600 BC. Hélène Borna Ahlkvist, who has studied these houses in detail, makes several important observations that are interesting in a discussion concerning the possible reuse of timber (2002).

From her analysis she draw a similar conclusion to Arnoldussen, namely, that long-houses were used for a period of about 75 years and that the abandoned houses were replaced by new houses built close to the old ones (Borna Ahlkvist 2002: 85–86). Moreover, among the houses at Pryssgården she identifies three groups of houses with very similar layouts, where the houses in each group presumably were built in sequence (2002: 37–42).

The houses in each group are approximately the same length and the roof-bearing constructions are identical in width. Moreover, the diameters of the roof-bearing post-holes are very similar among the houses in each group. According to Borna Ahlkvist, this indicates that the choice of timber dimensions was similar among the long-houses that replaced each other on the farm (Borna Ahlkvist 2002: 37–42).

In her interpretation the similarities in the width of the nave, the length of the house and the post-hole dimensions reflect tradition and knowledge inherited through generations belonging to the same family. Because the older people taught the younger people to build houses, the younger houses in many respects ended up similar to the older houses.

Perhaps the similarities are too striking to be merely explained by tradition.

In each of the three groups of houses Borna Ahlkvist identified there are buildings that are strikingly similar when it comes to the position of the roof-bearing posts. In the first group House 187 and House 194 are identical. In the second group four out of six (or seven according to interpretation) trestles in houses 210 and 211 have identical positions, while in the third group five out of seven (or nine according to interpretation) trestles in House 172 and House 175 have identical positions (information from plans in Borna Ahlkvist 1998).

A way of explaining these coherences is to assume that substantial parts of the roof-bearing construction – except the roof-bearing posts – were reused. Such a reuse of, for example, beams connecting roof-bearing posts, and long beams connecting trestles, would certainly constrain the options for elaborating the house construction.

The houses could be slightly changed but the basic layout including the width of the nave was decided by the people who erected the first house on the farm. From this perspective the similarities between houses on the same farm are probably the combined result of constraints placed upon the house-building process by the reuse of timber and an inherited tradition of construction techniques.

To discuss the probable reuse of timber in more detail we will now turn our eyes to another major Bronze Age site – the excavation at *Lilla Köpinge* in southernmost Sweden.

Lilla Köpinge in southernmost Sweden

The excavation at Lilla Köpinge outside the town of Ystad in southernmost Sweden took place in 1979. It was part of a rescue excavation but the results were later integrated in an extensive research project involving many different disciplines (Berglund *et al.* 1991) and the excavation was discussed in research publications and in a doctoral thesis by the excavator Sten Tesch (1992, 1993). Recently Anna Gröhn has made a re-analysis of the houses in the area around Lilla Köpinge, where she underlines the striking similarities in house constructions, not only between different sites but also among houses on a particular site (2003).

In two trenches – covering 2,600 m² and 16,800 m² respectively – a settlement dating from the Late Neolithic to the Iron Age was uncovered. In total nine houses were dated to the Bronze Age. The distance between the two concentrations of three and six houses respectively is approximately 150 metres. In addition to proper long-houses three outbuildings were recorded (Tesch 1993: 83–87, 89; Gröhn 2003: 212–215).

The long-houses were 12.3–14.5 metres long and 6.4–8.0 metres wide. The houses were well preserved, with both roof-bearing post-holes and most of the wall post-holes represented. The houses had four to five trestles and all of them were divided into two sections separated by an entrance room. The western part, having a wider interval between the trestles, and occasionally a hearth, was interpreted as a living area; the eastern part where the trestles were placed closer together was interpreted as a storeroom.

Several ^{14}C samples were taken from the houses, dating the Bronze Age settlement to the time span 1350–850 BC (Tesch 1993: 89). The three houses documented in one of the trenches (B13) presumably represent a farm where the long-houses were rebuilt in almost the same position. The site is not delimited to the east, so there might be hitherto undocumented buildings outside the trench. In the other trench (B14) six houses were documented. As has been proposed earlier, the nine houses in the two trenches probably represent 2–4 farms (Gröhn 2003: 311; Artursson 2009: 144).

There are certain phenomena in common between House I, House II and House VIII, situated in the same trench (B14), which indicate they were part of the same farm. They all had a pit along the southern wall of the living section that was filled with clay. A similar kind of pit was also found in House III in B13 but not in the other houses in trench B14. The pits might have been the bottom part of clay-lined storage pits (Gröhn 2003: 275) or, as suggested by the excavator, pits used for storing clay when making pots (Tesch 1993: 87).

Yet another feature is common to House I, House II and House VIII; they have an identical, or very similar, roof-bearing construction. In House VIII and House II the roof-bearing construction made of four trestles is identical. In House I, which is longer and has five trestles, the roof-bearing construction between the first and fourth trestles (counted from the west) is identical to those in House VIII and House II.

These two features – the pit with clay and the identical or very similar roof-bearing construction – differentiate House I, House II and House VIII from the other three houses in trench B14 (House III, House V and House VI). From these observations I suggest that House I, House II and House VIII represent a farm unit with a sequence of three houses built by the same family. An alternative view is that, since these houses are similar, they were built at the same time by the same person or rather group of persons.

However, it is only the roof-bearing construction that is identical while the outer walls are differently laid out in the three houses. The roof-bearing construction seems to be perfectly designed for only one of the houses, namely, House VIII. In this house the roof-bearing construction is centrally positioned inside the outer walls, while House II is asymmetrical designed with an extended gable in the east, and House I is slightly skewed along the axis (Figure 33).

These observations make it more likely that the houses were built in a sequence one after each other, but that the upper part of the original roof-bearing construction was reused when new houses were built. In the following we will continue this line of thought and see whether it is possible to make it a firmer conclusion.

The upper part of the roof-bearing construction is defined here as the roof-bearing construction except for the roof-bearing posts; i.e. crossbeam, high beam, gable head, wall head, connecting beam, the corner head and the door head and rafters (definitions from Draiby 1991).

House VIII is supposed to be the oldest house and was 12.2 metres long and 6.6–7.0 metres wide. It had four trestles and there was an entrance between the second and third trestle from the west. The posts were doubled in the western living section. The house is very symmetrical with straight gables. The distance between the roof-bearing posts and the outer wall is about 1.5 metres in all directions. Because of these qualities the house has been reconstructed in full scale at *Ekehagens forntidsby* outside Falköping in western Sweden (Tesch 1993: cover photo).

House II was 13.1 metres long and 6.6–7.0 metres wide. The distance between the roof-bearing construction and the longer outer walls is similar to that in house VIII, i.e. roughly 1.5 metre. But at the gables the walls are set at a greater distance from the trestles. In the western living area the difference is very tiny, only about 0.1–0.2 metres, but in the eastern part the outer wall was made round instead of straight. The maximum distance between the trestle and the eastern gable wall is 2.5 metres. This change in construction increased the area in the eastern store section by about 5 m², expanding the total area of the house from 83 m² to 88 m².

Assuming that upper part of the old roof-bearing construction from House VIII was reused, making a round gable instead of a straight one was the simplest way to increase the space. The only part of the upper roof-bearing construction that had to be renewed was the eastern gable rafters. However, the house acquired an asymmetrical exterior with one gable rounded instead of straight and the roof being at different angles at the gables.

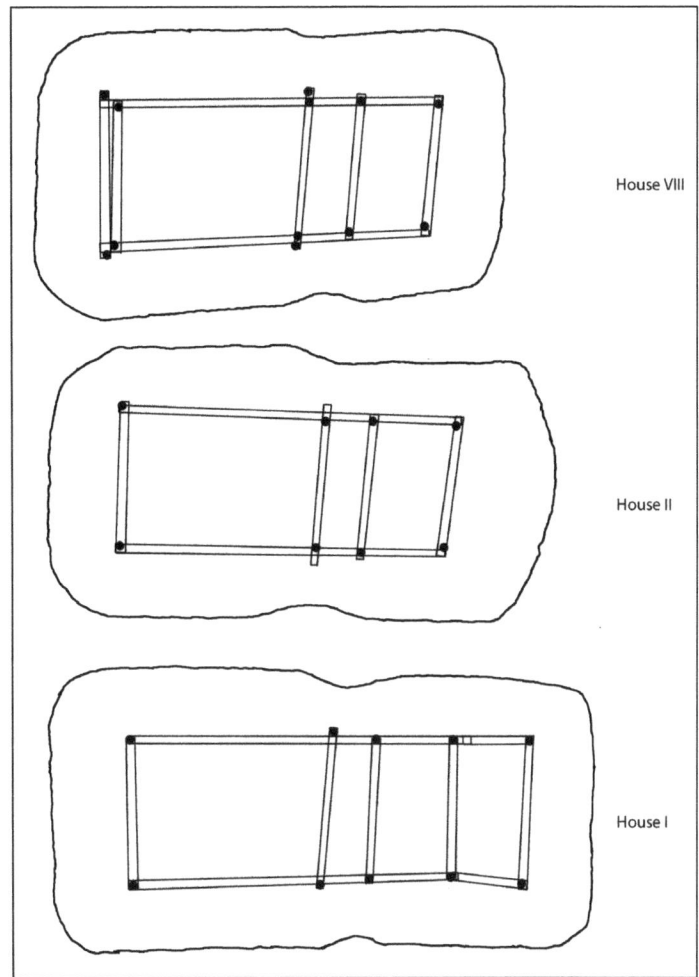

Figure 33. Plan of House I, House II and House VIII at Lilla Köpinge, Sweden (Trench B14), reproduced with an identical roof-bearing construction. The outer walls are reconstructed from the outer wall posts and the roof-bearing construction is inferred from the roof-bearing posts. Information from Tesch 1993: 168–169

Table 17. ^{14}C dates from Houses I, II and VIII in trench B14 at Lilla Köpinge. Data from Tesch 1993: 89

Lab No.	House No.	Feature	BP	±	One sigma (cal.)
St-7231	VIII	Wall post-hole	2905	160	1385–900 BC
St-7451	II	Door post-hole	2835	90	1157–957 BC
St 7452	II	Roof-bearing post-hole	2770	145	1121–810 BC
Lu-2455	II	Roof-bearing post-hole	2780	90	1036–833 BC
Lu-2462	II	Wall post-hole	2700	90	927–803 BC
St-7306	I	Wall post-hole door	2720	140	1020–800 BC

House I was 14 metres long and 6.9 to 7.2 metres wide. The roof-bearing construction between the first and fourth trestle (counting from the west) is identical to house VIII and I. However, this house had five trestles instead of four and the roof-bearing construction is extended by about 2 metres in comparison to House II and House VIII. The distance between the roof-bearing construction and the outer wall is roughly 1.5–1.6 metre, i.e. very similar to the measurements in House II and House VIII. Both the gables in this house are straight as in House VIII.

House I is slightly broader and the southern outer wall is not straight but bends somewhat outwards in the part of the house that is added in comparison to House VIII and House I. This is a kind of phenomenon that could occur if one adds a beam to an old roof-bearing construction and tries to maximize the width of the house.

The analysis of the position of the post-holes in the three houses therefore speaks in favour of timber reuse, or more precisely a reuse of the upper part of the roof-bearing construction on two occasions.

There are six ^{14}C dates from the three houses which are presented in table 17. The ^{14}C dates do not contradict the suggested house sequence – i.e. House VIII being the first followed by House II and then House I.

A WIDER PERSPECTIVE

Timber as history

When combining an estimated lifetime of about 75 years for an individual long-house, with the reuse of substantial parts of the timber construction, new perspectives are opened up on how to regard the biographies of long-houses. From these observations it is difficult to keep up with the idea of houses being abandoned at short and regular intervals and new houses being built from scratch. Instead we get a picture where houses stood on the same place for perhaps 75 years, and during this time several major repairs must have taken place. For example, if the roof was thatched with straw it was probably renewed every 25 years.

Following this line of thought we will get a building composed of material at differing ages. If we imagine a Bronze Age house built 50 years ago it could be composed of newly laid thatch, the material in the walls could be from time of the construction of the house 50 years ago, while much of the roof-bearing construction could be made from reused timer with an age of perhaps 200 years.

Probably timber was not randomly cut down but came from carefully planned woods, where coppicing favoured the availability of timber with different ages and diameters. To judge from the Lockarp example – when constructing a house, one was dependent upon decisions and woodland planning going back some 75 years in time. Considering both the woodland management and the reuse of timber, the life history of timber and trees could exceed the life history of the single house by centuries.

Based on a reading of anthropological literature, Fokke Gerritsen argues that there was an intimate relationship between the life cycle of the household and the life cycle of the individual house. These two processes ran parallel; when a new couple was married a new house was build at a new location. As the household expanded the house was repaired, extended and outbuildings were added, and finally at the death of the head of the household, the house was left to collapse or was burnt down (Gerritsen 1999).

In this perspective there is a very strong link between the family and the house. However, accepting that long-houses could stand for 50–100 years and accepting the results from the analysis of the Lilla Köpinge houses, there seem to have been other ways of ordering the relationship between a house and households.

Instead of an intimate relationship between individual houses and certain families we will get a strong relationship between houses and genealogy. Houses could have been constructed by timber cut down by relatives living 2–3 generations back, who in turn were dependent upon the woodland management of preceding generations. The house might contain timbers that were made by ancestors living ten generations ago; if so, history was inscribed in the house. Thus the house became a palimpsest of different times represented by wooden details of various ages that were joined together in the house construction (Olivier 1999).

Timber and architecture

Another perspective opened up by the recognition of timber reuse is that we get a slightly different understanding of variation among houses. For example, Anna Gröhn, in her discussion about the houses from Lilla Köpinge regards the exact measurements, straight lines and conformity between the posts in House II and House VIII as a sign of good architectural knowledge in comparison to House I (Gröhn 2004).

If we consider timber reuse, houses with asymmetrical layouts like House I from Lilla Köpinge in the first place reflect constraints placed upon the builders by the timber they used. In those cases where timber was used from various sources – like reused timber from different houses or a combination of reused timber and newly produced timbers – the idea of how to build the house was limited by the measurements and character of the individual pieces of timber available.

Thinking in these terms, houses with oblique walls, very asymmetrical plans, or tilted roof-bearing constructions to a very little extent reflect the knowledge and skills of the people making the houses, but instead give us hints about the limitations in the timber available at the construction site.

Prerequisites of timber reuse

It should be stressed that there is nothing regular in the reuse of timber. Reuse of timber most certainly occurred at some places and on some occasions in prehistory, but the temporal and spatial variation is evident. In some areas building material was easily available and on those occasions the reuse of timber was probably of less significance.

Reuse of timber requires timber of good quality that could be extracted from an abandoned house. This is not possible, for example, if the house was burnt down. A survey of burnt-down Iron Age houses from Denmark reveal that about 5% of some 1,100 registered house sites were burnt house sites. A majority of these are from north-west Denmark and were excavated in the early 20th century, when larger *Calluna* heathlands covered this part of the country (Christensen *et al.* 2007: 58–60). Probably many burnt house sites have been destroyed by modern agriculture and are underrepresented in the plains areas of northern Europe.

In some periods and areas there was certainly a shortage of timber. In the coastal areas of north-west Denmark the combination of high groundwater level close to the coast and raised sea beds, covering the houses, remains with sand has created very good preservation conditions (Bech 1997).

At the excavations of Bronze Age houses at *Bjerre* in north-west Denmark, the lower parts of the house posts were sometimes preserved. It could be noticed that not only oak had been used as roof-bearing posts, as might be expected, but that also alder and willow had been used. This clearly hints at a shortage of wood, as supported by other evidence. From pollen analysis we know that heathland was common already during the Bronze Age in this part of Denmark, and as a result of this turf had partially replaced wood as fuel at *Bjerre* already in the Bronze Age (Bech 1997). In such conditions one can expect the reuse of timber to be a necessity.

Chapter 5.
LINNAEUS' LANDSCAPE

Carl Linnaeus, the father of taxonomy, known in Sweden as Carl von Linné, was born on 23 May 1707 on the farm of Råshult, located close the city of Älmhult, at the Småland Uplands. Throughout his career Linnaeus kept to the creationist view of biological origins, which stipulated that studying nature reveals evidence for the creative powers and mysterious orderliness of God.

> But he did more than that: He treasured the diversity of nature for its own sake, not just for its theological edi-fication, and he hungered to embrace every possible bit of it within his own mind. He believed that humankind should discover, name, count, understand, and appreciate every kind of creature on Earth (Quammen 2007).

In this respect he was similar to the biologist Charles Darwin or the archaeologist Oscar Montelius living slightly more than a hundred years later; they all counted more on themselves than on God when setting out to collect and organize a huge bulk of data, in order to understand the ordering of the world and the position of humans both in their present time and in history.

According to Linnaeus himself, his long-life interest in nature and flowers came out of the garden his father, the vicar, laid out at the estate of Råshult. If we can trust Linnaeus, the Råshult garden had more herbaceous plants than any other garden in the province of Småland and it influenced Linnaeus a great deal, helping him to establish a lifelong love of flowers and botany.

In 2009 it was decided that the estate of Råshult would become a national park. This was a logical outcome of a process starting long before, aiming to preserve the ancient landscape around the birthplace of Linnaeus. Coming here today is a strange experience of entering a timeless landscape shaped by generations of humans. Here we find the clearance cairns scattered around in the fields with pollarded limes growing in them. The meadows are cleaned every spring and the leaves from last year are carefully burnt. The small and stony fields are ploughed by a single horse making its way between cairns and boulders.

The features of this landscape and the agricultural practices carried out here thus resembles the kinds of prehistoric landscapes that archaeologists and palaeoecologists reconstruct from various sources such as excavations and pollen analysis. Linnaeus' landscape, as it appears at Råshult, or at any other similar place, is therefore a fine starting point if one wishes to reflect upon some of the phenomena touched upon in this study; like the logistics of landscapes, the never-ending landscape change and the social significance of trees.

THE LOGISTICS OF LANDSCAPES

Any traditionally run farm, whether Linnaeus' birthplace or a prehistoric settlement, has to solve some basic problems by using the land it is in command of. The crops must be sown and harvested, the cattle taken care of and the fodder be collected and stored for the winter-time. In northern Europe the short summer gives the agricultural year a special rhythm, since very much of the outside work such as sowing, harvesting and fodder collection has to be done during a very limited period of time in the warm part of the year.

How these activities are carried out changes through history: the location and structure of fields, meadows and grazing land differs from time to time. Nevertheless, since the introduction of stone-cleared fields, which in this region appeared during the Early Bronze Age, these activities are presumed to have taken place inside a limited space not too far away from the dwelling houses of the farm.

At Råshult the production of fodder from lime leaves and twigs is still integrated in the management of the meadows. The pollarded trees are not randomly spread out in the meadows but grow in clearance cairns, proving these meadows to be former fields. The combination of pollarded trees and clearance cairns is logical. The stones cleared from the fields were placed at boulders and around trees to minimize the loss of cultivated land.

A similar phenomenon also occurs in prehistoric contexts. Here too, the clearance cairns are repeatedly placed on boulders to minimize the loss of cultivated land. The clearance cairns are generally set on a distance of five to ten metres. This pattern is something one would expect to occur if people are working together and throwing the stones by hand in an ordered way. But it is also a distance that works well for the ordering of pollarded trees in a field. These observations, taken together with the evidence from the charred remains of trees found below clearance cairns, indicate that clearance cairns were connected to pollarded and coppiced trees not only in the 18th century landscape but also in prehistory (Chapter 3).

At the same time as these supposed similarities are stressed, one must emphasize that the land was organized differently in prehistory. Palaeoecological research carried out at the estate of Råshult – including pollen analysis at two different places – has revealed a very clear pattern indicating the infield-outfield system still at use at Råshult has only existed from c. 1100 AD. The infield was primarily used for cultivation of crops in permanent fields and fodder production in meadows, while the outfield was primarily used for grazing and occasionally for slash-and-burn agriculture. The activities carried out in the infields, such as haymaking and subsequent pasture, helped to establish a very high biodiversity in the infields, which has no counterpart in the outfield (Lindbladh and Bradshaw 1995, 1998).

A characteristic Råshult shares with many other farms in the South Swedish Uplands is that outside the fields cultivated today (the historical infields), there are large areas with clearance cairns located in today's wooded outland, testifying that earlier agriculture was organized in a different way, involving larger areas of stone-cleared fields.

A possible explanation for this pattern is that there was a kind of rotation partly identified by the use of trees for fodder, fuel and carpentry. If pollarded and coppiced trees grew in the fields, as was discussed at length in Chapter 4, the rotation cycle could have ranged from couple of years to perhaps eight years, as is known from eastern England during the Middle Ages. When the trees were coppiced and pollarded, the tree canopy was reduced, thus facilitating crop production. Thereafter the trees could regenerate again during a couple of years before it was again time for pollarding and coppicing. During this period the fields were used for grazing.

The main difference between Linnaeus' landscape and the Bronze Age landscape would be that in prehistory cereal production, fodder production and grazing were conducted on the same piece of land, but in different years, while in the 18th century landscape the production of fodder by haymaking and pollarding was separated from the production of cereals in the fields. If these suggestions are accepted, one could say that the basic components of Linnaeus' 18th century landscape were present already in the Bronze Age, although they were organized differently. Today's heritage at Råshult thus cannot be understood without a sense of how humans have shaped this environment for millennia by clearing their fields of stones and by pollarding trees.

NEVER-ENDING LANDSCAPE CHANGE

From the pollen analysis it seems clear that a farm was established at Råshult c. 1100 AD. The establishment of the farm at Råshult was part of a larger north-European trend with population increase and colonization of former uninhabited areas from the 8th to the 13th century (Lagerås 2007: 26–31). The period preceding the farm establishment is somewhat confusing. There are no archaeological records indicating that the estate was settled during the Late Iron Age; two stone settings, the three cup-mark sites and areas with fossil arable fields in the vicinity of today's farm demonstrate that a farm existed here during the Bronze Age and Early Iron Age.

The establishment of the infield-outland system around 1100 AD started a process that led to a clear division of the landscape. The infield became more open, characterized by fields and meadows with pollarded trees, while the outlands gradually turned into a conifer forest. This process started in the Middle Ages when pine increased as a result of slash-and-burn agriculture occasionally carried out in the outlands (Lindbladh and Bradshaw 1998).

However, even more dramatic changes were to come; in the 20th century the rural areas of the South Swedish Uplands were depopulated as a result of urbanization and vast areas, mainly the former outlands, were turned into spruce plantations (Lagerås 2007: 153–163). A detailed study of an area about 70 kilometres south-west of Råshult demonstrated that in 1865 forest covered 23% of the land, while today there is a forest cover of 73% and 85% of this is coniferous (Lagerås 2007: 162). The study reveals a pattern that can be found in many parts of the South Swedish Uplands.

The forest history of the South Swedish Uplands thus runs parallel to what happened in New England; this was touched upon at the very beginning of the book. As in New England, the people of the South Swedish Uplands left their homes in the countryside and found employment in the urban factories. As a result of emigration, many of these people ended up in the great cities of North America like New York and Chicago. Their houses back home in Sweden were often left to decay and forest was planted in their stone-cleared fields.

As in New England, wildlife returned as a result of forestation. The landscape of Linnaeus was a landscape shaped and characterized by domestic animals. Today cattle have become rare but instead wildlife has returned as a result of forestation. Elk, roe deer, wild boar and fox today walk the forest of the South Swedish Uplands, something that was unthinkable only a century ago.

From an archaeological point of view this dramatic development is just one of many cultural landscape changes taking place in history. Judging by the archaeological record, landscape change might be defined as taking place at three different levels. The first level is the planned and regular rotation of cultivated fields inside a given area, as was discussed in Chapter 3. Trees were left to regenerate for a couple of years to produce fuel and fodder and were then cut back, and the land was used for agriculture, and after harvest the trees were left to regenerate again.

The second level concerns the life-history of the farm, which was discussed in Chapter 4. Evidence from various regions speaks in favour of single farms having existed during a period of perhaps 100–200 years before being abandoned or moved. These changes were primarily caused by factors occurring on a family level – farms were split, moved or abandoned as a result of population growth, or through death by catastrophes such as famines or wars.

The third level deals with dramatic changes involving larger landscapes. When studying the long-term development of agricultural landscape from the evidence of dated clearance cairns, or the settlement history in largely excavated areas, it is clear that at certain points in history landscapes were reorganized on a large scale. This happened, for example, at Hamneda around 100 AD when the formerly only extensively used woodlands were permanently settled. In these and other examples people left their old farms and moved into new areas, or they moved short distances and relocated their farms in new positions in the very same landscape (Björhem and Skoglund 2009).

Even though landscapes changed at regular intervals, there was a mosaic of various landscapes existing at the same time. In this study we have paid attention to the changes in the cultural landscape that occurred at Hamneda in the south-west Swedish Uplands around 100 AD when the moraine hills were turned into cultivated fields characterized by the growth of birch and hazel. Another important change took place a couple of centuries earlier, from the 9th to the 6th century BC in the Växjö area, in the southern central part of the South Swedish Uplands. In this area the low-lying areas close to alder swamps were cultivated and former grasslands were turned into *Calluna* heathland by fire. In chapter 3 these phenomenon were interpreted as intensified landscape use as a result of population increase.

An important conclusion to be drawn from these two examples is that the timing and phasing of prehistoric cultural landscape varied from area to area, and from region to region, depending upon the preceding settlement history of the region in question. Two inhabited and settled areas in the South Swedish Uplands could operate in two very different cultural landscape settings at the same time. Taking a bird's eye view of the South Swedish Upland landscape in the Early Iron Age, this means that in the region discussed there were simultaneously areas with fresh grasslands and a surplus of different kinds of trees and areas shaped by fire where *Calluna* heathland existed. This situation is very different from today, when the landscape in this region is homogenized over vast areas and dominated by either arable fields or spruce plantations.

However, in some villages the deciduous trees in the infields have been favoured for practical, aesthetic and cultural reasons (Mikusiński 2003). The few former infield areas that for some reason have not been part of the general trends of spruce plantation or rational agriculture are an important biological and cultural heritage to protect.

The deciduous and sometimes pollarded trees of traditionally managed inland areas are the last examples of the central position of trees in agriculture – a phenomenon going back more than 3,000 years in time. From the Late Bronze Age onwards pollarded and coppiced trees have presumably been an integral part of agricultural systems, not only in the Småland Uplands but in many parts of Europe.

The preservation of surviving pollarded trees is not only important for natural conservation, but it is also vital from the point of view of cultural heritage. If these types of landscapes are not preserved we will lose the possibility of studying the interplay and changing roles of material culture practices and biology in a landscape setting. The situation requires urgent attention, because until now the practice of pollarding and coppicing has prevailed in remote and hilly parts of Europe, but the situation is rapidly changing as a result of the ongoing urbanization and mechanization of agriculture (Emanuelsson 1996).

THE SOCIAL SIGNIFICANCE OF TREES

The naming and use of different species takes place in a social setting. The young Carl von Linnaeus was taught the names of the species at Råshult by his father. From his early childhood Linnaeus had a passionate interest in learning the names of different species. His father prevented him from learning new ones if he didn't remember the ones he already had learnt. As a result, his interest in names was even exaggerated.

One of the main achievements of Linnaeus was the establishment of a hierarchical structure of classification based upon observable characteristics and the introduction of a two-word naming system. He adopted the binomial nomenclature, using only the genus name and the specific name or epithet which together forms the species name. In this way he simplified and standardized the naming process and laid the foundation for a modern view of species classification.

In the 18th century the knowledge of species was put into the service of society. It was a shared goal for various kinds of academics to increase and facilitate the production of valuable materials inside the borders of the national state. On his travels all over Sweden Linnaeus often combined the descriptions of species and landscapes with ideas about how nature could be used in the service of humankind.

This is similar to what is argued in length in the book *The Social Life of Trees*; the combination of social and economic values cannot be separated when studying trees (Rival 1998). The combination is also noticeable in the Bronze Age, when trees took on both ritual and practical dimensions.

The tree had a ritual significance in the Bronze Age, as discussed in Chapter 2. Together with other important bearers of symbolic information the tree played an important role in helping to explain how the sun could move across the firmament. The ordering of these features on some of the bronze items might even be a justification for proposing that the tree had a central role to play in myths explaining the creation of humans out of chaos.

However, judging by the rock carvings in Bohuslän, these ideas did not grow out of nothing, but were firmly rooted in the domestic sphere where individual trees played a significant role. Especially the collecting of leaf fodder seems to have been important as an inspiration for rituals. In the Bronze Age the collecting of leaf fodder from high and shredded trees and the making of leaf-stacks were ritualized. This is by no means arbitrary since the shredded trees provided fodder for the cattle and thereby helped to preserve the foundations for life.

Linnaeus invented the binomial nomenclature, but anthropology teaches us that trees can be named on very different principles. In traditional Balinese societies the coconut tree is given different kinds of names depending upon its age but the names all refer to the social sphere; in contrast, the forest trees all have names referring to the wild sphere. These different naming principles testify to the importance of the coconut tree as a domesticated tree (Giambelli 1998: 134–135). A similar situation might have occurred in Bronze Age Scandinavia, where certain trees were treated differently depending upon their domestic use.

From this perspective it is interesting to note that the rock art and the species-identified charcoal seem to reveal two quite different kinds of stories. From the identified charcoal we get an idea of how low-growing trees and bushes, such as birch, alder and hazel, were taken to the settlements in large quantities to be used for fuel. These trees are almost absent in the rock art. In contrast, in the rock art we can study the use of tall trees – assumed to be primarily ashes, limes, maples and elms – and how they were used for rituals in connection with the practice of pollarding.

Speaking in general terms, we might be able to identify two categories of trees in the Bronze Age which differed from each other in both practical and ritual terms. On the one hand there were the less ritualized low-growing trees that were cut down in large numbers and primarily used for fuel and carpentry. On the other hand there were the high-growing, to a great extent ritualized, trees like ash, lime, maple and elm, which were primarily used for fodder production and carpentry. In comparison to the low-growing trees, the high-growing trees were more seldom cut down and were often left in the landscape for many generations.

In between these two categories is the oak. The oak is a high-growing tree but it was often cut down and it occurs in almost all contexts. It was used for fuel, carpentry, house building, but less often pollarded. It occurs in many practical and ritual contexts; moreover, the oak is also the most prominent bearer of history among all trees.

If left to live in the landscape, oak trees grow old, but when houses were built, younger oak trees, ranging from perhaps 25 to 75 years, were chosen with a diameter that was appropriate for the particular piece of timber needed. On the other hand, by being used as timber, these quite young oak trees could gain a significant age as "dead trees".

Its has been argued in this study that individual timbers now and then were reused in new buildings during the Bronze Age. Thereby the close link between not only generations of different houses but also different human generations was confirmed and established. The different colours and textures of newly made and old reused timber, reminded the people living in the houses of the passing of time and their place in the chain of people now dead and people to come.

In our modern world we still need those places that enable a sense of continuity over centuries and even millennia. From this perspective, places like Råshult matter – here we can study the interplay between humans and trees and how this relationship have emerged throughout millennia of continuous, but frequently changing, landscape use.

References

AKERET, Ö., HAAS, J.N., LEUZINGE, U. and JACOMET, S. 1999. Plant macrofossils and pollen in goat/sheep faeces from the Neolithic lake-shore settlement Arbon Bleiche 3, Switzerland. *The Holocene* 9 (2): 175–182.

ALMGREN, O. 1927. *Hällristningar och kultbruk: Bidrag till belysning av de nordiska bronsåldersristningarnas innebörd.* Stockholm: Wahlström & Widstrand.

ALTHIN, C.-A. 1945. *Studien zu den Bronzezeitlichen Felszeichnungen von Skåne I–II.* Lund: Gleerup; Copenhagen: Munksgaard.

ANATI, E. 1994. *Valcamonica rock art: A new history for Europe.* Capo di Ponte: Edizioni del Centro.

ANDERSSON, R. 2005. *Historical land-use information from culturally modified trees.* Umeå: Sveriges lantbruksuniversitet.

ANDERSEN, S.H., LIND, B. and CRUMLIN PEDERSEN, O. 1991. *Slusegårdgravpladsen: Bornholm fra 1. årh. f. til 5. årh. e.v.t. 3, Gravformer og gravskikke, bådgravene.* Højbjerg: Jysk Arkæologisk Selskab.

ANDERSSON, T. 1997. *Högsbyn i Tisselskogs Socken.* Arkeologisk rapport 2 från Vitlyckemuseet. Dals Långed: Ask och Embla.

ANDRÉN, A. 1993. Doors to other worlds: Scandinavian death rituals in Gotlandic perspectives. *Journal of European Archaeology,* 1993 (1): 33–56.

ANDRÉN, A. 2004. I skuggan av Yggdrasil: Trädet mellan idé och realitet i nordisk tradition. In Andrén, A., Jennbert, K. and Raudvere, C. (eds.), *Ordning mot kaos: Studier av nordisk förkristen kosmologi,* 389–430. Lund: Nordic Academic Press.

ANDRÉN, A. 2006. Skandinavisk religion i tid och rum. In Andrén, A. and Carelli, P. (eds.), *Odens öga: Mellan människor och makter i det förkristna Norden,* 3–45. Helsingborg: Dunkers kulturhus.

ANDRÉN, A., JENNBERT, K. and RAUDVERE, C. 2006. Old Norse Religion: Some problems and prospects. In Andrén, A, Jennbert, K. and Raudvere, C. (eds.), *Old Norse religion in long-term perspectives: Origins, changes, and interactions. An international conference in Lund, Sweden, June 3–7, 2004,* 11–14. Lund: Nordic Academic Press.

ARCINI, C., HÖST, E. and SVANBERG, F. 2007. Gravar, bålplatser och två bronsåldersfamiljer i Gualöv: Studier av en gravmiljö. In Artursson, M. (ed.), *Vägar till Vætland: En bronsåldersbygd i nordöstra Skåne 2300–500 f. Kr.,* Stockholm: Riksantikvarieämbetets förlag.

ARNOLDUSSEN, S. 2008. *A living landscape: Bronze Age settlement sites in the Dutch river area, c. 2000–800 BC.* Leiden: Sidestone Press.

ARTURSSON, M. 2009. *Bebyggelse och samhällsstruktur: Södra och mellersta Skandinavien under senneolitikum och bronsålder 2300–500 f. Kr.* Göteborg: Göteborgs universitet.

ÅSTRAND, J. 2009. *Flathällamon – ett kulturlandskap från bronsålder invid Växjö flygplats.* Växjö: Smålands museum.

BAILLIE, M. 1995. *A slice through time: Dendrochronology and precision dating.* London: Routledge.

BALTZER, L. 1881–1908. *Hällristningar från Bohuslän I–II. Tecknade och utgifne af L. Baltzer med förord af Victor Rydberg.* Göteborg.

BARRET, S.W. SWETNAM, T.W. and BAKER, W.L. 2005. Indian fire use: Deflating the legend. *Fire Management Today* Vol. 65: (3): 31–33.

BARTHOLIN, T. 1996. Den neolitiske hasselskov. In Slotte, H. and Göransson, H. (eds.), *Lövtäkt och stubbskottsbruk: Människans förändring av landskapet – boskapsskötsel och åkerbruk med hjälp av skog 1–2.* Stockholm: Kungliga Skogs- och Lantbruksakademien.

BARTHOLIN, T. and BERGLUND, B.E. 1992. The prehistoric landscape in the Köpinge area – a reconstruction based on charcoal analysis. In Larsson, L., Callmer, J. and Stjernquist, B. (eds.), *The archaeology of the cultural landscape: field work and research in a south Swedish rural region,* 345–358. Stockholm: Almqvist & Wiksell International.

BECH, J.-H. 1997. Bronze Age settlements on raised seabeds at Bjerre, Thy, NW-Jutland. In Dobiat, C. and Leidorf, K. (eds.), *Forschungen zur Bronzezeitlichen Besiedlung in Nord- und Mitteleuropa*. Espelkamp: Verlag Marie Leidorf.

BELL, C.M. 1992. *Ritual theory, ritual practice*. New York: Oxford University Press.

BENGTSSON, L. and OLSSON, C. 2000. *Världsarvsområdets centrala del och Grebbestad*. Tanumshede: Vitlycke museum.

BERGENDORFF, C. and EMANUELSSON, U. 1996. History and traces of coppicing and pollarding in Scania, south Sweden. In Slotte, H. and Göransson, H. (eds.). *Lövtäkt och stubbskottsbruk: Människans förändring av landskapet – boskapsskötsel och åkerbruk med hjälp av skog*, 235–312. Stockholm: Skogs- och lantbruksakademien.

BERGLUND, B.E. (ed.) 1991. *The cultural landscape during 6000 years in southern Sweden: The Ystad Project*. Copenhagen: Munksgaard.

BERNANDOS, D., FOSTER, D., MOTZKIN, G. and CARDOZA, J. 2004. Wildlife dynamics in the changing New England landscape. In Foster, David R. and Aber, John D. (eds.), *Forests in time: The environmental consequences of 1,000 years of change in New England*, 142–168. New Haven: Yale University Press.

BERNTSSON, A. 2005. *Två män i en båt: Om människans relation till havet i bronsåldern*. Lund: University of Lund, Institute of Archaeology.

BJÖRHEM, N. and MAGNUSSON STAAF, B. 2006. *Långhuslandskapet: En studie av bebyggelse och samhälle från stenålder till järnålder*. Malmö: Malmö kulturmiljö.

BJÖRHEM, N. and MAGNUSSON STAAF, B. 2006. *Långhuslandskapet: En studie av bebyggelse och samhälle från stenålder till järnålder*. Malmö: Malmö Kulturmiljö.

BJÖRHEM, N. and SÄFVESTAD. 1989. *Fosie IV: Byggnadstradition och bosättningsmönster under senneolitikum*. Malmö: Malmö museer.

BJÖRHEM, N. and SÄFVESTAD, U. 1993. *Fosie IV: bebyggelsen under brons- och järnålder*. Lund: Lund University.

BJÖRHEM, N. and SKOGLUND, P. 2009. Kulturlandskapets kontinuitet – platser, gårdar och vägar i ett långtidsperspektiv. In Högberg, A., Nilsson, B. and Skoglund, P. (eds.), *Gården i landskapet: Tre bebyggelsearkeologiska studier*, 17–98. Malmö: Malmö Museer.

BLOCH, M. 1998. Why Trees, Too, Are Good to Think With: towards an Anthropology of the meaning of Life. In Rival. L. (ed.). *The social life of trees: Anthropological perspectives on tree symbolism*. Oxford: Berg.

BORGEGÅRD, S.-O. 1996. Exposé över lövtäkt i tryckta dokument från 1700-talet till modern tid. In Slotte, H. and Göransson, H. (eds.), *Lövtäkt och stubbskottsbruk: Människans förändring av landskapet – boskapsskötsel och åkerbruk med hjälp av skog 1–2*, 121–157. Stockholm: Kungliga Skogs- och Lantbruksakademien.

BORNA-AHLKVIST, H. 1998. *Pryssgården: Från stenålder till medeltid. Arkeologisk slutundersökning RAÄ 166 och 167, Östra Eneby socken, Norrköpings kommun, Östergötland*. Linköping: Riksantikvarieämbetet, Byrån för arkeologiska undersökningar.

BORNA AHLKVIST, H. 2002. *Hällristarnas hem: Gårdsbebyggelse och struktur i Pryssgården under bronsålder*. Lund: Lund University.

BOYE, V. 1896. *Fund af Egekister fra Bronzealderen i Danmark. Et monografisk Bidrag til Belysning af Bronzealderens Kultur*. Copenhagen.

BRADLEY, R. 1997. *Rock art and the prehistory of Atlantic Europe: Signing the land*. London: Routledge.

BRADLEY, R. 2005. *Ritual and domestic life in prehistoric Europe*. London: Routledge.

BRENNAND, M. and TAYLOR, M. 2003. The survey and excavation of a Bronze Age timber circle at Holme-next-the-Sea, Norfolk, 1989–9. *Proceedings of the Prehistoric Society* 69: 1–84.

BREUNING-MADSEN, H., HOLST, M.K. and RASMUSSEN, M. 2001. The chemical environment in a burial mound shortly after construction – An archaeological-pedological experiment. *Journal of Archaeological Science* 28: 691–697.

BREUNING-MADSEN, H., HOLST, M.K., RASMUSSEN, M and ELBERLING, B. 2003. Preservation within log coffins before and after barrow construction. *Journal of Archaeological Science* 30: 343–350.

BRONGERS, J.A. and WOLTERING, P.J. 1978. *De prehistorie van Nederland: Economisch-technologisch*. Haarlem: Fibula-Van Dishoeck.

BUCKLEY, D.G., HEDGES, J.D. and BROWN, N. 2001. Excavations at a Neolithic Cursus, Springfield, Essex, 1979–85. *Proceedings of the Prehistoric Society* 67: 101–162.

BURENHULT, G. 1980. *Götalands hällristningar I*. Tjörnarp: Eget förlag.

CAPELLE, T. 1986. Schiffssetzungen. *Praehistorische Zeitschrift*, 61: 1–63.

CHRISTENSEN, L.B., JENSEN, S.E., LUND JOHANSEN, A.L., JOHANSEN, P.R. and LERAGER, S. 2007. House 1 – experimental fire and archaeological excavation. In Rasmussen, M. and Lund Hansen, U. (eds.), *Iron Age houses in flames: Testing house reconstructions at Lejre*, 41–133. Lejre: Historical-Archaeological Experimental Centre.

CLARK, P. (ed.) 2004. *The Dover Bronze Age Boat*. Swindon: English Heritage.

COLES, B. and COLES, J. 1986. *Sweet track to Glastonbury – the Somerset Levels in Prehistory*. New York: Thames and Hudson.

COLES, J.M. 2000. *Patterns in a rocky land: Rock carvings in south-west Uppland, Sweden* 1. Uppsala: Department of Archaeology and Ancient History.

CRONBERG, C., SKOGLUND, P. and TORSTENS-DOTTER ÅHLIN, I. 2000. Järnåldersgården och åkern. In Lagerås, P. (ed.), *Arkeologi och paleoekologi i sydvästra Småland: Tio artiklar från Hamnedaprojektet*, 145–165. Lund: Avdelningen för arkeologiska undersökningar, Riksantikvarieämbetet.

CRONON, W. 1983. *Changes in the land: Indians, colonists, and the ecology of New England.* New York: Hill and Wang.

CRUMLEY, C.L. (ed.) 1994. *Historical ecology: Cultural knowledge and changing landscapes.* Santa Fe: School of American Research Press.

CRUMLIN-PEDERSEN, O. and BONDESEN, E. 2002. *The Skuldelev ships.* I. Topography, archaeology, history, conservation and display. Roskilde: Viking Ship Museum.

CRUMLIN-PEDERSEN, O. and TRAKADAS, A. 2003. (eds.), *Hjortspring: A Pre-Roman Iron Age warship in context.* Roskilde: Viking Ship Museum in Roskilde.

DAY. G.M. 1953. The Indian as an ecological factor in the northeastern forest. *Ecology* Vol. 34 (2): 329–346.

DJUPEDAL, R. and BROHOLM, H.C. 1953. Marcus Schnabel og Bronzealderfundet fra Grevensvænge. *Aarbøger for Nordisk Oldkyndighed og Historie* 1952: 5–59. Copenhagen.

DRAIBY, B. 1991. Studier i jernalderens husbygning: Rekonstruktion af et langhus fra ældre romersk jernalder. In Madsen, B. (ed.), *Eksperimentel arkæologi.* Lejre: Historisk-Arkæologisk Forsøgscenter.

ELIASSON, L. and KISHONTI, I. 2003. *Öresundsförbindelsen: Lockarp 7B: rapport över arkeologisk slutundersökning.* Malmö: Malmö Kulturmiljö.

ELIASSON, L. and KISHONTI, I. 2007. *Det funktionella landskapet: Naturvetenskapliga analyser ur ett arkeologiskt perspektiv.* Malmö: Malmö Kulturmiljö.

EMANUELSSON, U. 1996. Lövängar och liknande markanvändningstyper i Europa. In Slotte, H. and Göransson, H. (eds.). *Lövtäkt och stubbskottsbruk: Människans förändring av landskapet – boskapsskötsel och åkerbruk med hjälp av skog*, 215–234. Stockholm: Skogs- och lantbruksakademien.

EMANUELSSON, U. 2002. *Det skånska kulturlandskapet.* Lund: Naturskyddsföreningen. i Skåne.

EMANUELSSON, M. 2003. *Skogens biologiska kulturarv: Att tillvarata föränderliga kulturvärden.* Stockholm: Riksantikvarieämbetets förlag.

EVANS, C., POLLARD, J. and KNIGHT, M. 1999. Life in woods: Tree-throws, settlement and forest cognition. *Oxford Journal of Archaeology* 18: 241–254.

FOSTER, D.R. and ABER, J.D. (eds.), *Forests in time: The environmental consequences of 1,000 years of change in New England.* New Haven: Yale University Press.

FOSTER, D.R., MOTZKIN, G., O'KEEFE, J.O., BOOSE, E., ORWIG, D., FULLER, J. and HALL, B. 2004. The environmental and human history of New England. In Foster, D.R. and Aber, J.D. (eds.), *Forests in time: The environmental consequences of 1,000 years of change in New England*, 43–100. New Haven: Yale University Press.

FOSTER II, H.T., BLACK, B. and ABRAMS, M.D. 2004. A witness tree analysis the effect of Native American Indians on the Pre-European settlement forests in East-Central Alabama. *Human Ecology* 32 (1): 27–47.

FREDELL, Å. 2002. Hällbilden som förskriftligt fenomen – en ansats för nya tolkningar. In Goldhahn, J. (ed.), *Bilder av bronsålder: Ett seminarium om förhistorisk kommunikation. Rapport från ett seminarium på Vitlycke museum 19–22 oktober 2000*, 24–260. Stockholm: Almqvist & Wiksell International.

FREDELL, Å. 2003. *Bildbroar – figurativ bildkommunikation av ideologi och kosmologi under sydskandinavisk bronsålder och förromersk järnålder.* Göteborg: Institutionen för arkeologi, Göteborgs universitet.

FREDSJÖ, Å., NORDBLADH, J. and ROSVALL, J. 1975. *Hällristningar Kville härad i Bohuslän. Bottna socken.* Göteborg: Fornminnesföreningen i Göteborg i samarbete med Göteborgs arkeologiska museum.

FREDSJÖ, Å., NORDBLADH, J. and ROSVALL, J. 1981. *Hällristningar Kville härad i Bohuslän. Kville socken, Del 1–2.* Göteborg: Göteborgs fornminnesförening i samarbete med Göteborgs arkeologiska museum och Institutionen för arkeologi vid Göteborgs universitet.

GARNER, A. 2004. Living history: Trees and metaphors of identity in an English forest. *Journal of Material Culture* Vol. 9 (1): 87–100.

GELLING, P. and DAVIDSON, H. 1969. *The chariot of the sun and other rites and symbols of the Northern Bronze Age.* London: Dent.

GERRITSEN, F. 1999. The cultural biography of Iron Age houses and the long-term transformation of settlement patterns in the southern Netherlands. In Fabech, C. and Ringtved, J. (eds.), *Settlement and landscape: Proceedings of a conference in Århus, Denmark, May 4–7 1998*, 139–148. Højbjerg: Jutland Archaeological Society.

GIAMBELLI, R.A. 1998. The coconut, the body and the human being: Metaphors of life and growth in Nusa Penida and Bali. In Rival. L. (ed.). *The social life of trees: Anthropological perspectives on tree symbolism.* Oxford: Berg.

GIESECKE, T. 2004. *The Holocene spread of spruce in Scandinavia.* Uppsala: Uppsala University.

GIESECKE, T. and BENNETT. K.D. 2004. The Holocene spread of *Picea abies* (L.) Karst. in Fennoscandia and adjacent areas. *Journal of Biogeography* 31: 1523–1548.

GLOB, P.V. 1951. *Ard og plov i Nordens oldtid.* Aarhus: Universitetsforlaget.

GLOB, P.V. 1962. Kultbåde fra Danmarks Bronzealder. *KUML* 1961: 9–18.

GLOB, P.V. 1969. *Helleristninger i Danmark.* Copenhagen.

GOLDHAHN, J. 2005. *Från Sagaholm till Bredarör – hällbildsstudier 2000–2004.* Göteborg: Institutionen för arkeologi, Göteborgs universitet.

GOLDHAHN, J. 2006. *Hällbildsstudier i norra Europa: Trender och tradition under det nya millenniet.* Göteborg: Institutionen för arkeologi, Göteborgs universitet.

GOLDHAHN, J. 2007. *Dödens hand – en essä om brons- och hällsmed.* Göteborg: Institutionen för arkeologi och antikens kultur, Göteborgs universitet.

GOLDHAHN, J., FUGLESTVEDT, I. and JONES, A. (eds.), *Changing pictures: Rock art traditions and visions in Northern Europe.* Oxford: Oxbow Books.

GÖRANSSON, H. 1995. *Alvastra pile dwelling: Paleoethnobotanical studies.* Stockholm: Lund University Press.

GRÄSLUND, B. 1987. *The birth of prehistoric chronology: Dating methods and dating systems in nineteenth-century Scandinavian archaeology.* Cambridge: Cambridge University Press.

GREISMAN, A. 2009. *The role of fire and human impact in Holocene forest and landscape dynamics of the boreo-nemoral zone of Southern Sweden: A multiproxy study of two sites in the province of Småland.* Kalmar: Kalmar University.

GREISMAN, A, GAILLARD, M.-J. and SKOGLUND, P. In Greisman, A., Late Holocene climate, human impact, fire, forest dynamics and long-term *Calluna* heath history in southern Sweden. (submitted to *Vegetation History and Archaeobotany*.)

GREISMAN, A. and GAILLARD, M.-J. 2009. The role of climate variability and fire in early and mid Holocene forest dynamics of southern Sweden. *Journal of Quaternary Science* Vol. 24: 593–611.

GREN, L. 1989. Det småländska höglandets röjningsröseområden. *Arkeologi i Sverige 1986*, 73–95. 1986. Stockholm: Fornminnesavdelningen, Riksantikvarieämbetet.

GREN, L. 1996. Hackerörens landskap och extensivt jordbruk under bronsålder – äldre järnålder. In Slotte, H. and Göransson, H. (eds.), *Lövtäkt och stubbskottsbruk: Människans förändring av landskapet – boskapsskötsel och åkerbruk med hjälp av skog 1*, 371–408. Stockholm: Kungliga Skogs- och Lantbruksakademien.

GRÖHN, A. 2004. *Positioning the Bronze Age in social theory and research contexts.* Lund: Lund University.

HAAS, J.N., KARG, S. and RASMUSSEN, P. 1998. Beech Leaves and Twigs used as Winter Fodder: Examples from Historic and Prehistoric Times In Charles, M. Halstead, P. and Jones, G. (eds.), *Fodder archaeological, historical and ethnographic studies. Environmental Archaeology* 1: 81–86.

HÆGGSTRÖM, C.-A. 1996. Hamlade träd i konsten. In Slotte, H. and Göransson, H. (ed.), *Lövtäkt och stubbskottsbruk: Människans förändring av landskapet – boskapsskötsel och åkerbruk med hjälp av skog 1–2*, 159–185. Stockholm: Kungliga Skogs- och Lantbruksakademien.

HÆGGSTRÖM, C.-A. 1998. Pollard meadows: Multiple use of human-made nature. In Kirby, K. and Watkins, C. 1998. *The ecological history of European forests*, 33–41. Wallingford: CAB International.

HALSTEAD, P. and TIERNEY, J. 1998. Leafy hay: An ethnoarchaeological study in NW Greece. In Charles, M. Halstead, P., and Jones, G. (eds.), *Fodder archaeological, historical and ethnographic studies. Environmental Archaeology* 1: 71–80.

HARDING, A. F. 2000. *European societies in the Bronze Age.* Cambridge: Cambridge University Press.

HASTORF, C.A. and JOHANNESSEN, S. 1996. Understanding changing people/plant relationships in the Prehispanic Andes. In Preucel, R.W. and Hodder, I. *Contemporary archaeology in theory: A reader*, 61–78. Oxford: Blackwell.

HAYASHIDA, F.M. 2005. Archaeology, ecological history, and conservation. *Annual Review of Anthropology* Vol. 34: 43–65.

HILLAM, J. GROVES, C.M., BROWN, D.M., BAILLIE, M.G.L., COLES, J.M. and COLES, J.B. 1990. Dendrochronology of the English Neolithic. *Antiquity* 64: 210–220.

HÖGBERG, T. 1995. *Litsleby, Tegneby and Bro.* Arkeologisk rapport 1 från Vitlyckemuseet. Uddevalla: Bohusläns museum.

HÖGRELL, L. 2000. Backar och bygder: Om Hamneda sockens fasta fornlämningar. In Lagerås, P. (ed.), *Arkeologi och paleoekologi i sydvästra Småland: Tio artiklar från Hamnedaprojektet*, 85–111. Lund: Avdelningen för arkeologiska undersökningar, Riksantikvarieämbetet.

HUMPHREY, C. and LAIDLAW, J. 1994. *The archetypal actions of ritual: A theory of ritual illustrated by the Jain rite of worship.* Oxford: Clarendon Press.

HVASS, S. 1985. *Hodde: Et vestjysk landsbysamfund fra ældre jernalder.* Copenhagen: Akademisk forlag.

HYENSTRAND, Å. 1984. *Fasta fornlämningar och arkeologiska regioner.* Stockholm: Riksantikvarieämbetet.

HYGEN, A.-S. and BENGTSSON, L. 1999. *Hällristningar i gränsbygd: Bohuslän och Östfold.* Sävedalen: Varne förlag och Riksantikvarieämbetet.

IMER, L.M. 2004. Gotlandske billedsten – dateringen af Lindqvists gruppe C og D. *Aarbøger for nordisk Oldkyndighed og Historie* 2001: 47–112.

KARG, S. 1998. Winter and spring foddering of sheep/goat in the Bronze Age site of Fiavè-Carera,

Northern Italy. In Charles, M. Halstead, P., and Jones, G. *Fodder archaeological, historical and ethnographic studies. Environmental Archaeology* 1: 87–94.

KAUL, F. 1985. Priorsløkke – en befæstet jernaldersby fra ældre jernalder ved Horsens. In *Nationalmuseets arbejdsmark 1985*, 172–183. Copenhagen: The National Museum.

KAUL, F. 1988. *Da våbnene tav: Hjortspringfundet og dets baggrund*. Copenhagen: The National Museum.

KAUL, F. 1987. Sandagergård: A Late Bronze Age cultic building with rock engravings and menhirs from Northern Zealand, Denmark. *Acta Archaelogica* vol. 56: 31–54. Copenhagen.

KAUL, F. 1998a. *Ships on bronzes: A study in Bronze Age religion and iconography*. 1, *Text*. Copenhagen: National Museum.

KAUL, F. 1998b. *Ships on bronzes: A study in Bronze Age religion and iconography*. 2, *Catalogue of Danish finds*. Copenhagen: National Museum.

KAUL, F. 2000. Solsymbolet. *Skalk* 2000 (6): 28–31.

KAUL, F. 2003. The Hjortspring boat and ship iconography of the Bronze Age and Early Pre-Roman Iron Age. In Crumlin-Pedersen, O. and Trakadas, A. (eds.), *Hjortspring: A Pre-Roman Iron Age warship in context*, 187–207. Roskilde: The Viking Ship Museum in Roskilde.

KIRKBY, K.J. and WATKINS, C. 1998. *The ecological history of European forests*. Cambridge: CAB International.

KONIJNENDIJK, C.C. 2008. *The forest and the city: The cultural landscape of urban woodland*. Dordrecht: Springer Science.

KOPYTOFF, I. 1986. The cultural biography of things: Commodization as process. In Appadurai, Arjun (ed.), *The social life of things: Commodities in cultural perspective*, 64–90. Cambridge: Cambridge University Press.

KRISTIANSEN, K. and LARSSON, T.B. 2005. *The rise of Bronze Age society: Travels, transmissions and transformations*. Cambridge: Cambridge University Press.

LAGERÅS, P. 2000a. Järnålderns odlingssystem och landskapets långsiktiga förändring: Hamnedas röjningsröseområden i ett paleoekologiskt system. In Lagerås, P. (ed.), *Arkeologi och paleoekologi i sydvästra Småland: Tio artiklar från Hamnedaprojektet*, 167–229. Lund: Avdelningen för arkeologiska undersökningar, Riksantikvarieämbetet.

LAGERÅS, P. 2000b. (ed.) *Arkeologi och paleoekologi i sydvästra Småland: Tio artiklar från Hamnedaprojektet*. Lund: Avdelningen för arkeologiska undersökningar, Riksantikvarieämbetet.

LAGERÅS, P. 2002. Skog, slåtter och stenröjning. Paleoekologiska undersökningar i trakten av Stoby i norra Skåne. In Carlie, A. (ed.), *Skånska regioner*, 362–411. Stockholm: Riksantikvarieämbetet.

LAGERÅS, P. 2007. *The ecology of expansion and abandonment: Medieval and post-medieval agriculture and settlement in a landscape perspective*. Stockholm: National Heritage Board.

LAGERÅS, P. and BARTHOLIN, T. 2003. Fire and stone clearance in Iron Age agriculture – new insights inferred from the analysis of terrestial macroscopic charcoal in clearance cairns in Hamneda, southern Sweden. *Vegetation History and Archaeobotany* Vol. 12. (12) 83–92.

LARSSON, G. 2007. *Ship and society: Maritime ideology in late Iron Age Sweden*. Uppsala: Uppsala University.

LÉVI-STRAUSS, C. 1966. *The savage mind*. Chicago: University of Chicago Press.

LINDBLADH, M. and BRADSHAW, R. 1995. The development and demise of a medieval forest-meadow system at Linnaeus' birthplace in southern Sweden: Implications for conservation and forest history. *Vegetation History and Archaeobotany* 4: 153–160.

LINDBLADH, M. and BRADSHAW, R. 1998. The origin of present forest composition and pattern in southern Sweden. *Journal of Biogeography* 25: 463–477.

LINDBLADH, M. 2004. När granen kom till byn. *Svensk botanisk tidskrift* 98: 249–262.

LINDQVIST, S. 1941–42. *Gotlands Bildsteine* 1–2. Stockholm: Almqvist & Wiksell.

LING, J. 2008. *Elevated rock art: Towards a maritime understanding of Bronze Age rock art in northern Bohuslän, Sweden*. Göteborg: Göteborgs universitet.

LINSCOTT, K. 2007. *Medeltida tak: Bevarade takkonstruktioner i svenska medeltidskyrkor. Del 1, Rapport om kunskapsläget 2006*. Göteborg: Göteborgs universitet. Institutionen för kulturvård.

MALMER, M.P. 1981. *A chorological study of North European rock art*. Stockholm: Kungliga Vitterhets Historie och Antikvitetsakademien; Almqvist and Wiksell International.

MARLON, J.F., CUI, Q., GAILLARD, M.J., McWETHY, D. and WALSH, M. 2010. Humans and fire: Consequences of anthropogenic burning during the past 2 ka. *PAGES news*, Vol 18 (2) 80–82.

MARSTRANDER, S. 1963. *Østfolds jordbruksristninger* I–II. Oslo.

MIKUSIŃSKI, G., ANGELSTAM, P. and SPORRONG, U. 2003. Distribution of deciduous stands in villages located in coniferous forest landscapes in Sweden. *Ambio* 32 (8): 520–526.

MORPHY, H. 1991. *Ancestral connections: Art and an aboriginal system of knowledge*. Chicago: Univerity of Chicago Press.

MYRDAL, J. 1984. Elisenhof och järnålderns boskapsskötsel i Nordvästeuropa *Fornvännen* 79.

NILSSON, B. and SKOGLUND, P. 2000. To dwell in the centre of the world: On the life history of a gallery-grave in Småland, SE Sweden. *Lund Archaeological Rewiev* 6: 43–60.

NILSSON, L.E. 2005. *Hjul på hällar*. Sävedalen: Warne.

NORDÉN, A. 1925. *Östergötlands bronsålder*. Linköping: Carlssons bokhandel.

NORMAN P. 1989. Röjningsrösen och förhistoriska graver. *Arkeologi i Sverige 1986*: 97–109. Stockholm: Fornminnesavdelningen, Riksantikvarieämbetet.

NYLÉN, E. and LAMM, J.P. 2003. *Bildstenar*. Stockholm: Gidlunds.

OHLSON, M. and TRYTERUD, E. 1999. Long-term spruce forest continuity a challenge for a sustainable Scandinavian forestry. *Forest Ecology and Management* 124: 27–34.

OLDEBERG, A. 1933. *Det nordiska bronsåldersspännets historia med särskild hänsyn till dess gjuttekniska utformning i Sverige*. Stockholm.

OLIVIER, L.C. 1999. Duration, memory and the nature of the archaeological record. In Andersson, G. and Karlsson, H. (eds.), *Glyfer och arkeologiska rum – en vänbok till Jarl Nordbladh*, 529–533. Göteborg: Institutionen för arkeologi, Göteborgs universitet.

OLSSON, F, GAILLARD, M.-J., LEMDAHL, G, GREISMAN, A., LANOS, P., MARGUERIE, D., MARCOUX, N., WÄGLIND, J. and SKOGLUND, P. 2010. A continuous record of fire covering the last 10500 calendar years from southern Sweden – the role of climate and human activities. *Palaeogeography, Palaeoclimatology, Palaeoecology* Vol. 291: 128–141.

OLSSON, F. and LEMDAHL, G. 2009. A continuous Holocene beetle record from the site Stavsåkra, southern Sweden: implications for the last 10 600 years of forest and land use history. *Journal of Quaternary Science* Vol. 24: 612–626.

ÖSTLUND, L., BERGMAN, I. and ZACKRISSON, O. 2004. Trees for food – a 3000 year record of subarctic plant use. *Antiquity* 78: 278–286.

ÖSTLUND, L., ZACKRISSON, O. and HÖRNBERG, G. 2002. Trees on the border between nature and culture – Culturally modified trees in boreal Scandinavia. *Environmental History* 7(1): 48–68.

PENACK, J.J. 1993. *Die eisernen eisenzeitlichen Erntegeräte im freien Germanien*. Oxford: BAR Publishing.

PERINI, R. 1987. *Scavi archeologici nella zona palafitticola di Fiavè-Carera. P. 2, Campagne 1969–1976: Resti della cultura materiale: metallo-osso-liticalegno*. Trento: Servizio beni culturali della Provincia Autonoma di Trento.

PERINI, R. 1995. *Scavi archeologici nella zona palafitticola di Fiavè-Carera. P. 1, Campagne 1969–1976: Situazione dei depositi e dei resti strutturali*. Trento: Servizio beni culturali della Provincia Autonoma di Trento.

PRYOR, F. 2002. *Seahenge: A quest for life and death in Bronze Age Britain*. London: HarperCollins.

PYNE, S.J. 1982. *Fire in America: A cultural history of wildland and rural fire*. Seattle: University of Washington Press.

QUAMMEN, D. 2007. A passion for order. *National Geographic*. June 2007.

RACKHAM, O. 2003. *Ancient woodland: Its history, vegetation and uses in England*. Dalbeattie: Castlepoint Press.

RAIHLE, J. and HOMMAN, O. 2005. Medeltida timmerhus i Dalarna. *Dalarna*. 2005(75): 81–124.

RAMQVIST, P.H. 1997. *Inte bara väggar: Analys av bränd lera från järnåldern*. Umeå: Umeå University.

RANDSBORG, K. and CHRISTENSEN, K. 2006. Bronze Age oak-coffin graves. *Acta Archaeologica* 77: 1–246.

REGNELL, M. 2002. Arkeobotanisk analys. In Torstensdotter Åhlin, I., Skoglund, P., Cronberg, C., Gustafsson, P. and Högrell, L. *Boplatslämningar och röjningsrösen: Småland, Ljungby kommun, Hamneda socken, RAÄ 66, 67, 76 och 82: Arkeologisk undersökning*. Lund: Riksantikvarieämbetet/Växjö: Smålands museum.

REGNELL, M. 2003a. Charcoals from Uppåkra as indicators of leaf fodder. In Larsson, L. and Hårdh, B. (eds.), *Centrality – regionality: The social structure of southern Sweden during the Iron Age*, 105–115. Stockholm: Almqvist and Wiksell International.

REGNELL, M. 2003b. Växtmakrofossil från RAÄ 77. In Torstensdotter Åhlin, I., Skoglund, P. Engman, F. Jonsson, L. Lagerås, P. Linderoth, T. Regnell, M. and Svanberg, F. *Gravar, röjningsrösen och boplatslämningar: Småland, Ljungby kommun, Hamneda socken, RAÄ 77: Arkeologisk undersökning*. Lund: Riksantikvarieämbetet; Växjö Smålands museum.

RIVAL, L. 1998a. Trees, from symbols of life and regeneration to political artefacts. In Rival. L. 1998 (ed.), *The social life of trees: Anthropological perspectives on tree symbolism*. Oxford: Berg.

RIVAL. L. 1998b (ed.). *The social life of trees: Anthropological perspectives on tree symbolism*. Oxford: Berg.

SABATINI, S. 2007. House urns: *Study of a late Bronze Age trans-cultural phenomenon*. Göteborg: Göteborgs universitet.

SÄFVESTAD, U. 1995. Husforskning i Sverige 1950–1994. In Kyhlberg, O., Göthberg, H. and Vinberg, A. (eds.), (1995). *Hus & gård i det förurbana samhället. Artikeldel*, 11–22. Stockholm: Avdelningen för arkeologiska undersökningar, Riksantikvarieämbetet.

SCHOVSBO, P.O. 1987. *Oldtidens vogne i Norden: Arkæologiske undersøgelser af mose- og jordfundne vogndele af træ fra neolitikum til ældre middelalder*. Fredrikshavn: Odense University.

SHETELIG, H., FALK, H. and BRØGGER, A.W. (eds.) 1917–2006. *Osebergfundet*. Kristiania.

SILLASOO, Ü. 2006. Medieval plant depictions as a source for archaeobotanical research. *Vegetation History and Archaeobotany* 16: 61–70.

SKOGLUND, P. 2005. *Vardagens landskap: Lokala perspektiv på bronsålderns materiella kultur.* Lund: Lunds universitet.

SKOGLUND, P. 2006. *Hällristningar i Kronobergs län: Motiv, myter och dokumentation.* Lund: Institutionen för arkeologi och antikens historia, Lunds universitet.

SKOGLUND, P. 2007. Landscape, History and Monuments – A Material Culture Perspective. In Salisbury, R.S. and Keeler, D. (eds.), *Space – Archaeology's Final Frontier. An Intercontinental Approach*, 244–271.

SKOGLUND, P. 2008. The composite ard in Sweden – its introduction, geographical distribution and consequences for landscape management. In Goldhahn, J. (ed.), *Gropar och Monument: En vänbok till Dag Widholm*, 323–339. Kalmar: University of Kalmar.

SKOGLUND, P. 2009. Beyond chiefs and networks: Corporate strategies in Bronze Age Scandinavia. *Journal of Social Archaeology* 9 (2): 200–219.

SKOGLUND, P. in press. Social landscapes of Bronze Age Scandinavia. In Anfinset, N. and Wrigglesworth, M. (ed.), *Local societies, identities and responses: The Bronze Age in Northern Europe.* London: Equinox Publishing.

SKOGLUND, P. and LAGERÅS, P. 2002. En vendeltida vagn från södra Småland: Fyndet från Skirsnäs mosse i ny belysning. *Fornvännen* 97 (2): 73–86.

SKOGLUND, P. and SVENSSON, E. 2010. Discourses of nature conservation and heritage management in the past, present and future: Discussing heritage and sustainable development from Swedish experiences. *European Journal of Archaeology* Vol. 13:3: 368–385.

SLOTTE, H. 2000. *Lövtäkt i Sverige och på Åland: Metoder och påverkan på landskapet.* Sveriges lantbruksuniversitet. Uppsala.

Smålands museum Rapport 2003:60. Särskild arkeologisk undersökning. Hällkista samt fossil åkermark med gravar längs Rottnevägen RAÄ 175, 206 m fl Gårdsby socken Växjö kommun.

SPROCKHOFF, E. 1956. *Jungbronzezeitliche Hortfunde der Südzone des nordischen Kreises. Bd 1–2.* Mainz: Verlag des Römisch-germanischen Zentralmuseums.

STRYD, A.H and FEDDEMA, V. 1998. *Sacred cedar: The cultural and archaeological significance of culturally modified trees.* Vancouver: The David Suzuki Foundation.

TESCH, S. 1992. House, farm and village in the Köpinge area from the Neolithic to the early Middle Ages. In Larsson, L., Callmer, J. and Stjernquist, B. (eds.), *The archaeology of the cultural landscape: Field work and research in a south Swedish rural region*, 283–384. Stockholm: Almqvist and Wiksell International.

TESCH, S. 1993. *Houses, farmsteads, and long-term change: A regional study of prehistoric settlements in the Köpinge area, in Scania, southern Sweden.* Uppsala: Uppsala university.

THRANE, H. 1975. *Europæiske forbindelser: bidrag til studiet af fremmede forbindelser i Danmarks yngre broncealder (periode IV–V)* Copenhagen: Nationalmuseet.

TILLHAGEN, C.-H. 1995. *Skogarna och träden: Naturvård i gångna tider.* Stocholm: Carlsson.

TOLLIN, C. 1989. Röjningsrösen i södra Sverige. *Arkeologi i Sverige* 1986: 53–71. Stockholm: Fornminnesavdelningen, Riksantikvarieämbetet.

TORSTENSDOTTER ÅHLIN, I., SKOGLUND, P. ENGMAN, F. JONSSON, L. LAGERÅS, P. LINDEROTH, T. REGNELL, M. and SVANBERG, F. 2003. *Gravar, röjningsrösen och boplatslämningar: Småland, Ljungby kommun, Hamneda socken, RAÄ 77: Arkeologisk undersökning.* Lund: Riksantikvarieämbetet; Växjö Smålands museum.

WALKER, J. 1999. Late 12th and early 13th century aisled buildings: a comparison. *Vernacular Architecture* 30: 21–53.

WATKINS, C. 1998. *European Woods and Forests Studies in Cultural History.* Cambridge.

WATSON, C. 2005. Seahenge: An archaeological conundrum, English Heritage, Swindon.

WELINDER, S. 1992. *Människor och landskap.* Uppsala: Societas archaeologica Upsaliensis.

WELINDER, S. 2009. Den äldre järnålderns lilla landskap utanför Malmö. In Högberg, A., Nilsson, B. and Skoglund, P. (eds.), *Gården i landskapet: Tre bebyggelsearkeologiska studier,* 17–98. Malmöfynd 20. Malmö: Malmö museer.

WHITNEY, G.G. 1994. *From coastal wilderness to fruited plain: A history of environmental change in temperate North America, 1500 to the present.* Cambridge: Cambridge University Press.

WIDGREN, M. 1997. *Fossila landskap: En forskningsöversikt över odlingslandskapets utveckling från yngre bronsålder till tidig medeltid.* Kulturgeografiskt seminarium 97:1. Stockholm: Kulturgeografiska institutionen, Stockholms universitet.

WIDGREN, M. 2003. (ed). *Röjningsröseområdena på sydsvenska höglandet – arkeologiska, kulturgeografiska och vegetationshistoriska undersökningar.* Meddelanden från Kulturgeografiska institutionen vid Stockholms universitet 117.

WILLIAMS, M. 1989. *Americans and their forests: A historical geography.* Cambridge: Cambridge University Press.

WINTERBOURNE, A. 2004. *When the Norns have spoken: Time and fate in Germanic paganism.* Madison: Fairleigh Dickinson University Press.

WRIGHT, E. 1990. *The Ferriby boats: Seacraft of the Bronze Age.* London: Routledge.

ZACHRISSON, O. ÖSTLUND, L., KORHONEN, O. and BERGMAN, I. 2000. The ancient use of *Pinus Sylvestris* L. (Scots pine) inner bark by Sami people in northern Sweden related to cultural and ecological factors. *Vegetation History and Archaeobotany* 9: 99–109.

This book investigates the practical and ritual dimensions of trees and timber in the Bronze Age and Iron Age of Scandinavia. An important notion is that trees were designed to fit a broad variety of purposes – i.e. they were *culturally modified*. Trees were pollarded to gain leaf-fodder, woods were cut back at regular intervals to create timber and fuel with certain characteristics, and timber from older houses was reused in the building of new houses.

These arguments are developed through the study of a broad range of materials including rock-art images depicting trees, charcoal from archaeological contexts identified to species, and prehistoric long-houses. The archaeological remains are related to a wider discussion in anthropology and historical ecology concerning the multifaceted relationship between humans and trees.

www.ingramcontent.com/pod-product-compliance
Lightning Source LLC
Chambersburg PA
CBHW061550010526
44115CB00023B/2998